T0396073

Climate Change and
CIRCULAR ECONOMICS

Emerging Technologies and Materials in Thermal Engineering

Climate Change and CIRCULAR ECONOMICS

HUMAN SOCIETY AS A CLOSED THERMODYNAMIC SYSTEM

IONUT PURICA
National Institute for Economic Forecasting, Romanian Academy, Bucharest, Romania

Series editor
HAFIZ MUHAMMAD ALI

Elsevier
Radarweg 29, PO Box 211, 1000 AE Amsterdam, Netherlands
125 London Wall, London EC2Y 5AS, United Kingdom
50 Hampshire Street, 5th Floor, Cambridge, MA 02139, United States

ISBN: 978-0-443-29969-8

For information on all Elsevier publications visit our website at https://www.elsevier.com/books-and-journals

Publisher: Megan Ball
Acquisitions Editor: MFran Kennedy-Ellis
Editorial Project Manager: Ellie Barnett
Production Project Manager: Maria Bernard
Cover Designer: Mark Rogers

Typeset by TNQ Technologies

Dedication

To my father who gave me the wish for research and to Muni, my wife, who gave me the support and time to write this book.

Contents

About the author

Prof. Dr. ing. Dr. ec. Ionuţ Purica

Presently a Professor in the UHB, Executive Director of the Advisory Center for Energy and Environment, and President of the Romanian Association for Energy and Economy, he was also senior researcher in the Romanian Academy's Institute for Economic Forecasting. Prof. Purica was a member of the Advisory Group for Energy of the European Commission and of the WEC, Study Group for energy scenarios 2060, trilemma indicator, and counselor of the Minister of Research, Minister of Economy and previously the Minister of the Environment. He participated, among others, in the elaboration of the EU accession strategy for Romania and the energy (electricity and heat) strategy (for the Ministry of Economy and Trade) and did risk analysis and transaction structuring and project management in the World Bank (as a project officer) and with USEA, JBIC, MARSH, ITOCHU, MVV, etc. He is a corresponding member of the Academy of Romanian Scientists and he was also a President-State Secretary of the Romanian Agency for Nuclear Energy and Radioactive Waste.

Previously, he worked as a project officer for energy and infrastructure in the World Bank, in Romania and the Balkans (e.g. Energy Assessment in Kosovo 1999), to extend his expertise in engineering acquired as director for international projects of the Romanian Power Company RENEL and senior engineer managing a joint Atomic Energy of Canada Ltd-IMG-Bucharest quality engineering group for the manufacture of nuclear reactor components for the CANDU units in Romania. He also worked as an international researcher for ENEA Rome and as an associate researcher at ICTP Trieste.

Prof. Purica has authored books in his field of expertise published by Imperial College Press, Academic Press, etc. and published articles in journals like *Risk Analysis*, *IEEE Power Engineering Review*, *Foundations of*

Control Engineering, etc. He took his second Ph.D. in Economics (the first one in Nuclear Energy Engineering), and he is teaching a course in leadership and risk management to master of science students in the Politehnica University of Bucharest (international projects) and Hyperion University (UHB).

Foreword

It gives me immense pleasure to write this foreword on the new book titled *Climate Change and Circular Economics: Human Society as a Closed Thermodynamic System* published under the book series *Emerging Technologies and Materials in Thermal Engineering* by Elsevier. The author of this book Professor Ionut Purica is currently a Senior Researcher in Econophysics and a Member of the Academy of Romanian Scientists (AOSR). His educational background with two PhDs in Energy Systems and Economics, respectively, and enriched experience with the World Bank, ENEA Rome, ICTP Trieste, Italy and RENEL Romania in nonlinear models for energy systems development and economic decisions merit him a distinctive candidate to write on this very vital subject for the benefit of students, faculty members, engineers, researchers, and scientists working around the globe.

This book will serve as a unique ready reference with an access to the details of climate change in relation to circular economy with an emphasis on human society as a closed thermodynamic system. This book contains 10 distinctive chapters starting with an introduction (Chapter 1) on the concept of the role of human society in resource recovery by innovation determined through the irreversible thermodynamics method to resist and fight climate change. Chapter 2 presents insights into human society as an open dissipative system. Chapter 3 presents the concrete concept of circular economy based on the principal of irreversible thermodynamic and emphasizes on the need of turning waste into a useful asset with the help of resource management policy and new emerging technologies. Chapter 4 emphasizes the role of big data in analyzing climate change and corresponding risk mitigation—highlighted with a case study of natural gas networks of Italy and Romania. Chapters 5, 6 and 7 further advance into important economic indicators to deal with climate change and economics entropy, respectively. Chapter 8 details the information on "Gibbs Paradox" to describe the interconnection of separate systems with the help of multivalued modal logic, which gives a new perspective to deal with climate change and associated challenges. To enhance the complexity of sustainable development, in Chapter 9, a review is done of the 17 Sustainable Development Goals in the UN Strategy 2030. Finally, Chapter 10 draws conclusions for development that would go beyond fighting Nature to harmonizing with it.

In a nutshell, this book will provide a complete guide to the experts already dealing with climate change to advance their knowledge by adopting the new approach of circular economy to fight climate change. For beginners in the field of climate change, this book will prove to be a complete step-by-step guide through of various stages to develop and advance their knowledge to the next level.

March 2024

Dr. Hafiz Muhammad Ali
Associate Professor of Mechanical Engineering
King Fahd University of Petroleum and Minerals
Dhahran, Saudi Arabia
e-mail: Hafiz.ali@kfupm.edu.sa

Preface

Looking at the evolution of the GHG emissions expressed in CO_2 equivalent units of mass, one may see a continuous increase interrupted by short periods of decrease during the 'crises', be they financial, pandemic, energy, war, etc.

It seems that the interaction of the human society with the environment (Nature) is more complex than just greenhouse effect. The focus on just emissions is good as a parameter that may monitor the impact on climate change. But it creates a strange logical syllogism: climate change is produced by humans; fight climate change concludes in fighting ourselves. Up to a certain point, it may inspire action to diminish emissions through new technologies, but it eventually results in a too simple approach that neglects a wider vision on the process of interacting with Nature, thus, shifting priorities from technologies that manage resources and possibly transform waste into new resources, to the ones that are focused on emission reduction only.

A wider vision on the matter is attempted in this book by a thermodynamic approach that involves energy intensity, entropy, as well as the role of resource/waste management. This allows justifying the notion of circular economy that has penetrated in recent years, in a more comprehensive way. A model is done of a typical economy, using iThink/Stella software, that shows a dynamic behavior simulating energy intensity evolution, the Kuznets curve, emissions' evolution, and the fact that more classical energy resources we use, larger is the temperature increase, while more recovery leads to temperature decrease.

The evolution on long-term data of the energy intensity of selected economies opens the possibility to determine the impact on the environmental temperature resulting from the structural change implied by the technological evolution as well as to stress the importance of resource/waste management through innovation of new technologies aiming at changing liabilities into assets in relation to the environment. Moreover, the velocity of temperature increase is determined, allowing to compare with the environment time constant of recovery after the impact.

Since the amount of resource recycling and management is shown to have great importance in reducing the overall climate change effect, an analysis is done on resources, both from a recycling/availability and technology point of view and, also, an example is given of an energy balance—based decision to partition agricultural land between food for people and biofuel crops for cars in selected economies. Selected technologies are presented also, since technological resources are essential to shifting the importance of natural resources and to recycling waste. As a typical example, the introduction of electric vehicle technologies is shifting the importance of oil resources to the one of lithium resources. The evolution of battery technology may change the importance of lithium in the future. Nuclear and energy storage are having a strong penetration period enhanced by, respectively, the lack of emission and the need to compensate the high volatility of renewables.

On a different line, a big data analysis of the temperature and precipitations for a given country, as a good example to follow, shows that there is a real climate change effect of average temperature increase associated to an increase of the probability of extreme events such as drought and flood. Based on the results, a risk map may be generated at a country level that may help insurance and investment decisions for the various mitigation measures and for various regions where risks are manifesting themselves more intensely.

The dynamics of the economy/environment interaction leads to a more variate set of indicators and to a better understanding of such notions as resilience, sustainability, in the context of this approach. The need for more complex indicators is stressed by comparison with the ones we are using now that can be misleading in the perception of impacts and nonlinear behavior of societies. The cyclic evolution of technological innovation and its influence on the shortening of the Kondratiev economic cycles are also presented based on an analysis of the oscillatory evolution of the GDP components.

In order to have a consistent development, there is a need for specific investment schemes to implement adequate projects. A brief consideration is made on green investment schemes that may help the financing of sustainable projects.

To provide a set of selected information on the use of thermodynamic notions, a selection is done of various fragments from papers in the field. This includes a brief text generated using the recent ChatGPT AI software on the matter.

This book continues with a solution of the Gibbs 'paradox' that does not use the normal binary logic but a multivalued modal logic of possibilities that sheds a more profound light on the interaction of separate spaces versus their combination. This being a basis of economic and environmental spaces consideration and of the need for the decision makers to have complete information on the system.

The following chapter is focused on the recently defined Sustainable Development Goals by the United Nation and a case example is provided for their inclusion in a sustainable development strategy at a national level. This is a step forward in assessing the complexity of development and the needed indicators for specific decisions.

The final chapter is a set of considerations on the road ahead for reaching a sustainable future in a complex dynamic of the human society and environment.

In synthesis, the logic of the chapters is the following:

The first chapter presents the thermodynamic approach, stressing on the need to have a closed system in relation to the environment. Human society and Nature are two systems in interaction. The importance of having a closed Human society with regard to Nature is stressed in this chapter showing that it may lead to just an exchange of energy while the resources should be recycled thus returning possibly no material waste into the environment.

In this chapter, a model is built to simulate the behavior of the human economy and environment; here there are simulations, presented as graphics of evolution, various regimes of operation related to resources, energy, as well as natural ones, and to other parameters of interest. The result of these simulation regimes is pointing out the existence of limits beyond which the behavior changes, thus the existence of basins of behavior.

The second chapter introduces the method, based on the Gibbs equation in thermodynamics, of determining the environmental temperature increase induced by selected economies' evolution of energy intensity over long-term periods; the results are given in a table while the energy intensity is presented as a graph and digitized for calculations. The importance of resources recycling to temperature reduction results from this approach. Moreover, the rate of temperature increase is also determined, opening the way for comparing it to the time constant of environmental system recovery after an impact. From this point of view, the evolution of fast-developing emerging economies is seen in a new light in relation to the environment resilience.

The third chapter presents considerations on selected material resources and technological resources to mitigate climate change; there is also an analysis, based on energy conservation, of the decision to allocate land use between biofuel and food with examples from the United States of America, EU and China. The chapter underlines the need for a global resource and technological management if our interaction with the environment would be coherent.

The fourth chapter presents a case analysis of the climate change events risks maps based on big data series for temperature and precipitations having a geographical distribution for Romania. The risk maps are presented and an application is done for the impact on the Romanian gas network; also presented is an application for the risk map of the Italian gas network subject to earthquakes and land slide. The possibility to introduce an insurance policy to climate change risks is also determined based on the risk maps, as well as a better allocation of investments to mitigate, or adapt to, those risks.

The fifth chapter makes a description of indicators related to climate change in the greater context introduced in the book; there are no case studies here but various examples of integrated indicators that go beyond the usual GDP/cap or tCO_2/cap (i.e. two simple indicators). The chapter underlines the need for integrated indicators to reflect the evolution of the complex intercorrelation with the environment. An analysis of the oscillatory behavior of the GDP components is done as a support of the fact that economic innovation cycles contribute to shortening the Kondratiev cycles.

The sixth chapter makes a brief presentation of various green investment schemes designed to help implementation of sustainable projects in the seventh chapter.

The seventh chapter is gathering a selection of texts from the literature on the topic related to entropy in economics such that to give a feeling of the complexity of the topic. Given the recent use of ChatGPT, I have also included a text generated on the same topic (this text is of the order of magnitude of the other selected texts and should be taken as an example).

Chapter eight introduces an original approach on the Gibbs so called 'paradox' related to the interaction of two thermodynamic systems interconnected, being based on a modal logic approach (different from the usual binary logic) in terms of the information available to the decision maker. This chapter stresses the fact that for correct decisions one should not oversimplify the amount of information needed to the decision makers.

The ninth chapter presents the Sustainable Development Goals defined by the United Nations as a proof of the complexity of sustainable development process for the human society stressing again the need for a complex vision on this complex process.

The last chapter is making a synthesis of the conclusions and presents evolution scenarios for a sustainable future with insights into potential drawbacks delaying the reach of the final goal to have a coherent and durable relation between the human society and the environment.

References are given to each chapter.

Acknowledgments and credits

Credit: US Geological Survey
Credit: Visual Capitalist
Credit: William Polen USEA
Credit: Luminita Chivu INCE, Romania
Credit: Paolo Malanima, Magna Graecia University of Catanzaro

Acknowledgments

Glen Seaborg, former President of the US Atomic Energy Commission, for the information provided on nuclear that determined my choice of profession

Abdus Salam, former President of ICTP Trieste, for inspiring discussions on nonlinear systems

Giancarlo Pinchara, ENEA, for discussions on climate change and risk impacts.

Ioan Ursu, member of the Romanian Academy, for supervising my first doctorate in nuclear engineering

Aurel Iancu, member of the Romanian Academy, for supervising my second doctorate in economics

László Borbély, Minister, Department of Sustainable Development, for inspiring discussions on SDGs

CHAPTER ONE

Human society and nature interaction

[A living organism] … feeds upon negative entropy … Thus the device by which an organism maintains itself stationary at a fairly high level of orderliness (= fairly low level of entropy) really consists in continually sucking orderliness from its environment.

Erwin Schrodinger.

This first chapter presents the thermodynamic approach, stressing the need for a closed system in relation to the environment. Human society and nature are two interacting systems. The importance of having a closed human society with regard to nature is stressed in this chapter, showing that it may lead to just an exchange of energy while the resources should be recycled, thus possibly returning no material waste to the environment.

Is climate change real? An entropy view of the economy and environment intercorrelation. What are the risks of climate change phenomena?

Human society may be described as a system with a structure that evolves within an environment. It has the capacity to adapt the environment to its needs beyond the stage of systems that survive only if they can adapt to the environment.

Looking through the prism of the theory of thermodynamic systems, we can distinguish three types of systems: isolated, with a tendency to reach thermodynamic equilibrium; closed, which exchange only energy with the environment; and open, which also exchange substance with the environment. Closed and open systems can have equilibrium states far from the thermodynamic equilibrium involved in an organized structure.

We will try to prove that the rational solution for the evolution of human society is to move from the current state of an open system to one of closed states.

The maintenance of the organized structure of human society is tributary to the wind, rain and sun, and other fluctuating factors; and to raw, limited resources, a part of which become energy resources depending on the conversion processes that the development of science offers us: burning, fission, and fusion.

Climate Change and Circular Economics
ISBN: 978-0-443-29969-8
https://doi.org/10.1016/B978-0-443-29969-8.00012-4

Raw materials are transformed into useful products with the help of industrial technologies, as people need a portion of these products as the means of production, thus creating a name for themselves. Means of production are also needed to convert energy resources into useful energy—heat and electricity, part of which is used to realize technologies and included in industrial products, thus closing a second cycle specific to production processes. Following this, portions of useful energy and industrial products serve human consumption. Biologically maintaining the population requires food, which implies agriculture. The elimination of the fluctuating factors on which agriculture depends through industrial chemical technologies, irrigation, and mechanization tend to decouple agriculture from the environment, making it more dependent on solar energy and industrial processes.

Both the processes for transforming raw materials and those for consuming their products give rise to waste, of which a part is recoverable by specific technologies while the rest is not.

Nonrecoverable wastes are chemical, radioactive, or in the form of heat that cannot be used at the temperature of the environment.

The brief evolution of thermodynamics in relation to economic systems is provided in the Annex to this chapter.

1.1 Society as a dissipative open system

Considering human society and its relationship with the environment as described above results in a dissipative closed system in the sense of Ilya Prigogine.

Society takes raw materials and energy resources from the environment at low entropy, and after using them to maintain its organized structure and level of civilization, eliminates them as polluting environmental waste, thus creating dangerous conditions for societal development and irreversibly increasing environmental entropy.

Entropy characterizes the degree of utility of prime materials and energy; in other words, it increases with increasing disorder and is a measure of the degree of disorder in the case of an isolated system.

The production process is characterized by the combination of the two cycles referring to a hypercycle in the sense of M. Eigen, similar to describing the multiplication of a virus within a cell.

Maintaining the organized structure through the production process requires compensation for the increase in internal entropy by taking over negentropy from the environment and thereby disorganizing it. Our phrase "energetic processes" may be banal; likewise, from Georgescu Roegen's studies, we can associate entropy with economic processes, for which the phrase also becomes well defined.

A dissipative system is characterized by variations in entropy over time. If the system is in stable equilibrium and connected to thermodynamic equilibrium, the entropy growth rate should be minimal according to a theorem of I. Prigogine. This fact is not realized by our current society, which destroys the environment or, basically the same thing, exhibits an entropy growth rate that is rapidly increasing, causing environmental disorganization.

To determine an evolution that we should follow to avoid pollution that could lead to the disappearance of the current level of civilization or even the existence of human society, we must specify some ideas.

1.2 Closing processes in self-organizing cycles

Human society evolves not toward the renewing of order but by forming a new order that was not present under previous conditions.

So we consider the currently known universe with those 10^{80} particles that, according to the laws of physics, are associated with an immense number of combinations, and we find that these combinations have an evolutionary course of elementary lattices, atoms, molecules, biomolecules, cells, organs, and organisms associated with social and cultural structures.

We must note two equally important aspects of this evolution toward an increasing degree of ordering and thus reduced redundancy. At every transition from one level of ordering to another, for example, from biomolecules to cells or from humans to human society, we find ourselves facing a conflict resulting from the tendency to preserve an independent existence alongside the minimization of the risk of extinction. A biomolecule defends itself within a molecular environment at the risk of being destroyed through contact with other molecules. The association of biomolecules in an ordered structure, the cell, prioritizes certain biomolecules in relation to environmental molecules. This means that what changes by moving from one stage of organization to another is not only the appearance of a hyperstructure with a higher order but also the redefinition of the environment in which it evolves, develops, or disappears. The structure is formed by elements of the environment that, by the very appearance of the higher structure, acquire new values in relation to other environmental elements.

Human society gives humans a special position in relation to other mammals and other living things, whether animals or plants, embedded in an ecological niche where it is not humans but human society that evolves.

The appearance of a hyperstructure leads to the redefinition of the environment.

Even the evolution of the structure, through its transition to different stages of organization, can lead to a redefinition of the environment.

The second aspect we must consider and care about is the implicit concept of self-organization.

Self-organization, which has gained particular importance in recent decades for physical processes and systems (lasers), chemicals (for example, Belousov–Zhabotinsky reactions), and biological elements (Eigen's hypercycles), is a concept characterizing certain system behaviors.

Systems self-organize in their transition from complexity to complex order.

Self-organization results from the emergence of cyclical processes within the system. However, it does not mean self-maintenance, preserving its complex order through reactions initiated from within the system in order to face disturbances produced in interactions with the environment. Under this aspect, we can distinguish, as introduced by Varela, between the organization of the system, which is specific to it, and its structure, which can vary if the organization remains the same, depending on external conditions.

The organization of a social system is characterized by interactions between the system, considered as a unit, and its components, the people, considered as units. It is necessary to get used to the idea that an organized structure must be considered a unit at its level, although we consider its components as units at their level of organization. For this reason, H. Shimizu called the components "holon" and organized many of them under the term "holos" from the Greek word for "whole." He is currently working on a bioholonic project in Japan.

Through these interactions between components and their organized multitude, in the case of human society we are able to explain the stages of evolution, i.e., the situation in which, by assimilating a part of the environment, it becomes weakly coupled with the rest of the environment.

The key to this evolution is defined by the "closing" concept.

Any open system has a series of inputs and a series of outputs through which it interacts with other systems or the environment.

In the case of human society, the inputs are raw materials, energy resources, or information, and the outputs are the different categories of waste that pollute the environment. The transition from inputs to outputs occurs through sequences of processes within systems. Industrial processes transform raw materials into products, and through consumption or aging, they become chemical waste.

Closing these processes cyclically by transforming chemical waste into raw materials would lead to a new structure for both human society and the environment. The relationship between society and the environment is reduced to the consumption of raw materials only to ensure its growth and the acceptance of cyclical losses, which leads to a significant reduction in curtailment and dependence on the environment, giving it the qualities necessary for the existence of the components of society and people.

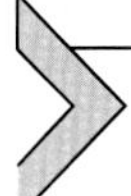

1.3 Evolution of human society toward a closed system

Evolution toward a closed system can be achieved through waste recovery; it is fully possible with the introduction of thermonuclear technologies and the use of the plasma torch.

Chemical waste can be decomposed by plasma torches at temperatures of tens of millions of degrees at the level of the component atoms. Then, through mass spectrometers, atoms are separated, concentrated, and reintroduced into the industrial circuit to absorb useful production.

However, industrial and energetic effort is required to produce the plasma torch. It produces radioactive waste and irrecoverable energy.

We can represent the characteristic cycles of production, energy, and waste through a simple, approximate model that allows us to analyze the behavior of human society as a closed system that exchanges only energy with the external environment.

We will consider the production of goods proportional to annual raw materials use, N, with the production effort measured according to the Cobb–Douglas function by the product $L^a K_p^b$, where L is labor forces, $K_p = e_k K$ are the means of production used to produce goods with energy consumed by the production process, $E_p = e_p E$ (e_p is the proportion of total energy used for production), and e_k is the proportion of total capital used to produce goods:

$$GDP = k_p * N * e_p * E * L\hat{}a * (e_k * K)\hat{}b$$

where k_p is a proportionality constant.

Production serves consumption, which is proportional to the population or labor force, c.L, for manufacturing the means of production, (K_p), energy, (K_e), and waste recovery (K_{rd}). These are proportional, respectively, to annual production, $K_p = e_k P$, to the energy produced annually, K_e $e_e E$, and to the waste that must be recycled annually, $K_{rd} = e_{rd}.E$.

The increase in the means of production must compensate for wear and tear, which is noted by the value $K_g = mK$.

Thus, a portion of production must be invested in the means of production, and we denote it by gP. Therefore, the annual increase in the means of production is

(1) $\delta K = gP\text{-}mK$

We analyze the case when the investment does not serve to increase but rather to maintain the means of production at a constant level, so, when $\delta K = 0$. In this case, the equation that gives us the variation over time of the annual production is

(2) dP/dt = kp*N*ep*ep*E*L̂a*(ek*K)̂b-(c*L + g*P)

Similarly, we can establish a balance equation for the annually produced energy, E:

(3) dE/dt = ke*R*L̂a*(ee*K)̂b-(el*L + ep*E + er*E + U)

The energy resources used annually are denoted by R; el is energy consumption per person for the population; and epE, erE are energy consumed for industrial production processes and, respectively, for waste recycling, depending on production levels; and U = energy loss during the conversion process.

The annual variation in the amount of waste is given by the equation

(4) dD/dt=(c*L + kd*N*ep*E*L̂a*(ek*K)̂b + K)-(krd*L̂a*(erd*K)̂ b + er*E*D + crd*D)

where it the production consumed cL is considered waste, but in the production process, a portion of the raw materials is lost and becomes waste with the coefficient k_d, the means of production replaced annually in mP are waste, and waste recycling implies k_{rd}, energy E, and effort proportional to the quantity of waste, D.

We thus have four equations that reflect the global behavior of the system and the associated model, as presented below.

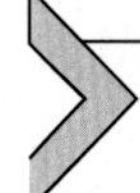

1.4 Model evolution 3

Top-Level Model:

D(t) = D(t − dt) + (Prod_deseuri − Reciclare_D) * dt {NON-NEGATIVE}

INIT D = 100

E(t) = E(t − dt) + (Prod_En − Cons_En) * dt {NON-NEGATIVE}

INIT E = 100

K(t) = K(t − dt) + (Invest_K − Pierderi_K) * dt {NON-NEGATIVE}

INIT K = 100

P(t) = P(t − dt) + (Prod_P − Cons_P) * dt {NON-NEGATIVE}

INIT P = 100

Cons_En = (el*L + ep*E + er*E + U) {UNIFLOW}

Cons_P = (c*L + g*P) {UNIFLOW}

Invest_K = g*P {UNIFLOW}

Pierderi_K = m*K {UNIFLOW}

Prod_deseuri = (c*L + kd*N*ep*E*L^a*(ek*K)^b + m*K) {UNIFLOW}

Prod_En = ke*R*L^a*(ee*K)^b {UNIFLOW}

Prod_P = kp*N*ep*ep*E*L^a*(ek*K)^b {UNIFLOW}

Reciclare_D = (krd*L^a*(er*K)^b + erd*E*D + crd*D*K) {UNIFLOW}

a = 0.3

b = 0.6

c = 0.2

Coef_CO_2 = GRAPH(TIME)

Points: (1.00, 99.3307149076), (2.00, 98.473284612), (3.00, 96.2), (4.00, 91.9), (5.00, 87.2), (6.00, 83.4), (7.00, 78.7), (8.00, 71.1), (9.00, 62.1), (10.00, 50.2), (11.00, 34.6), (12.00, 19.4), (13.00, 11.4)

crd = 0.5

ee = 0.06

ek = 0.9

el = 0.05

Emission_CO_2 = Coef_CO_2*Cons_En

En_int = E/(P + 0.01)

ep = 0.055

er = 0.1

erd = 0.03

g = 0.2

kd = 0.02

ke = 0.08

kp = 0.09

krd = 0.001

L = 100
m = 0.02
N = 20
R = 50
U = 10
{The model has 35 (35) variables (array expansion in parens).
In root model and 0 additional modules with 0 sectors.
Stocks: 4 (4); Flows: 8 (8); Converters: 23 (23)
Constants: 20 (20); Equations: 11 (11); Graphics: 1 (1)
}

Energy intensity evolves with a growth period followed by one of decline.

Emissions begin to decrease after a slight increase.

The Kuznets curve shows typical behavior i.e., after a certain value for GDP, emissions decrease.

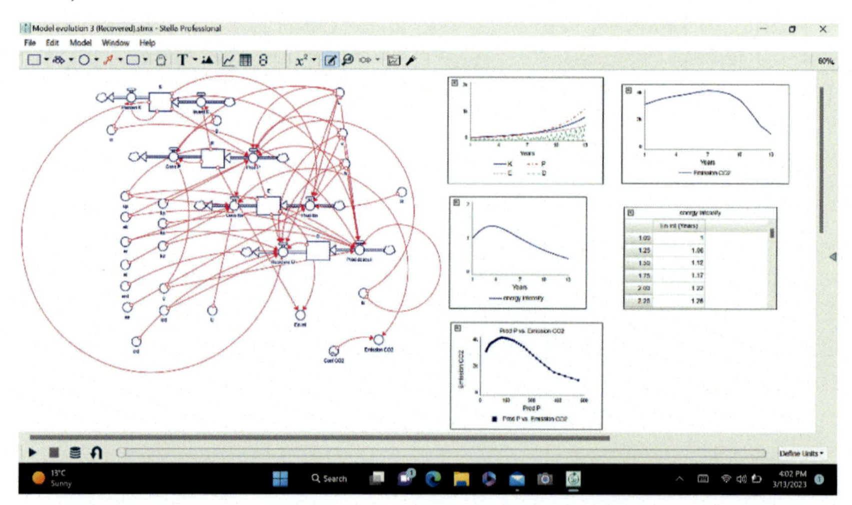

The model is generated using Stella for Business software.

1.5 Specific results

As an example of how to use the model, we take basic values for the parameters and consider some results related to finding changes in the behavior of variables resulting from changes in some parameters of interest.

Simulating the various regimes in the operation of a circular economy leads to some important conclusions related to changes in behavior. We have considered only the influence of the energy resources, R, and the natural resources, N, on the evolution of the primary model variables. Moreover, it is possible to explore behavior related to the level of capital destruction, m.

N = 20.

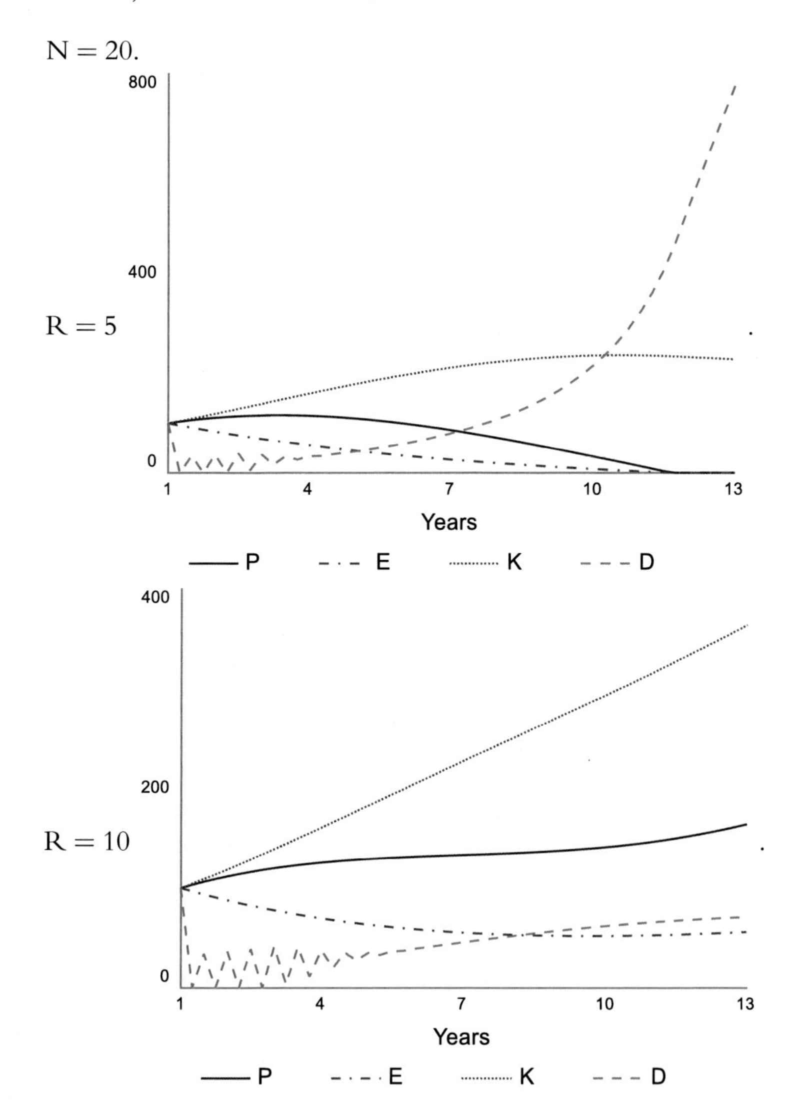

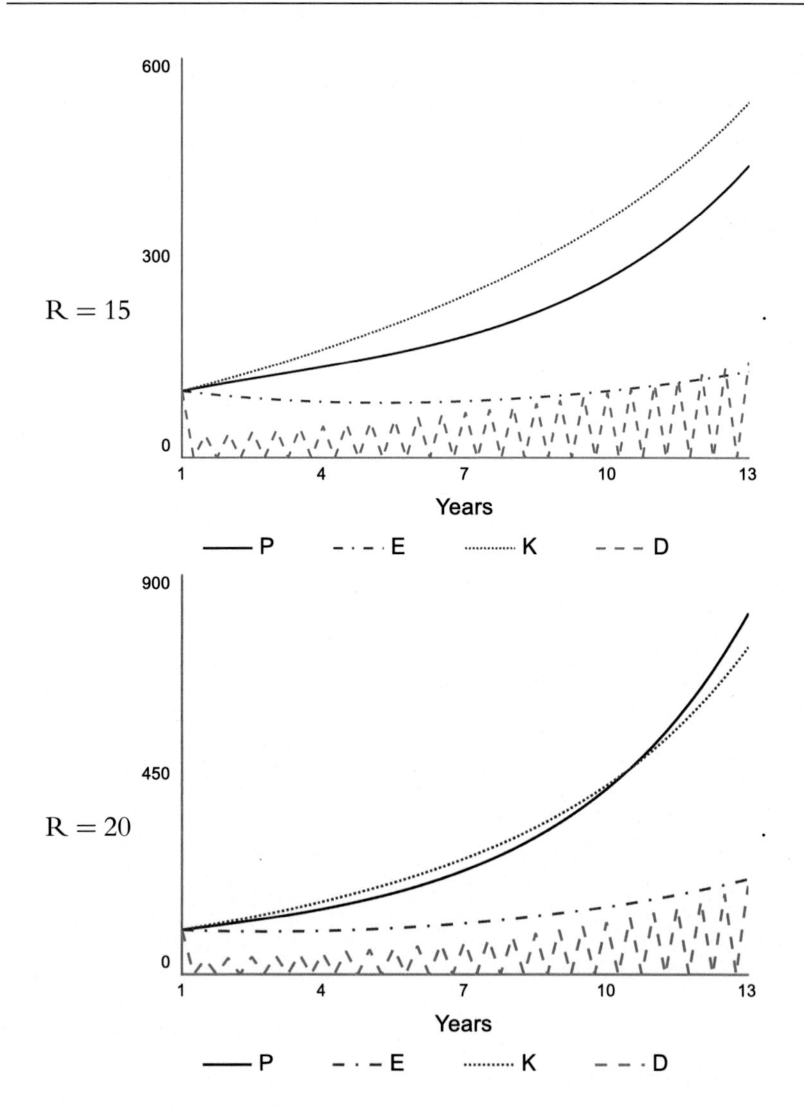
R = 15
600
300
0
1
4
7
10
13
Years
P
E
K
D
R = 20
900
450
0
1
4
7
10
13
Years
P
E
K
D

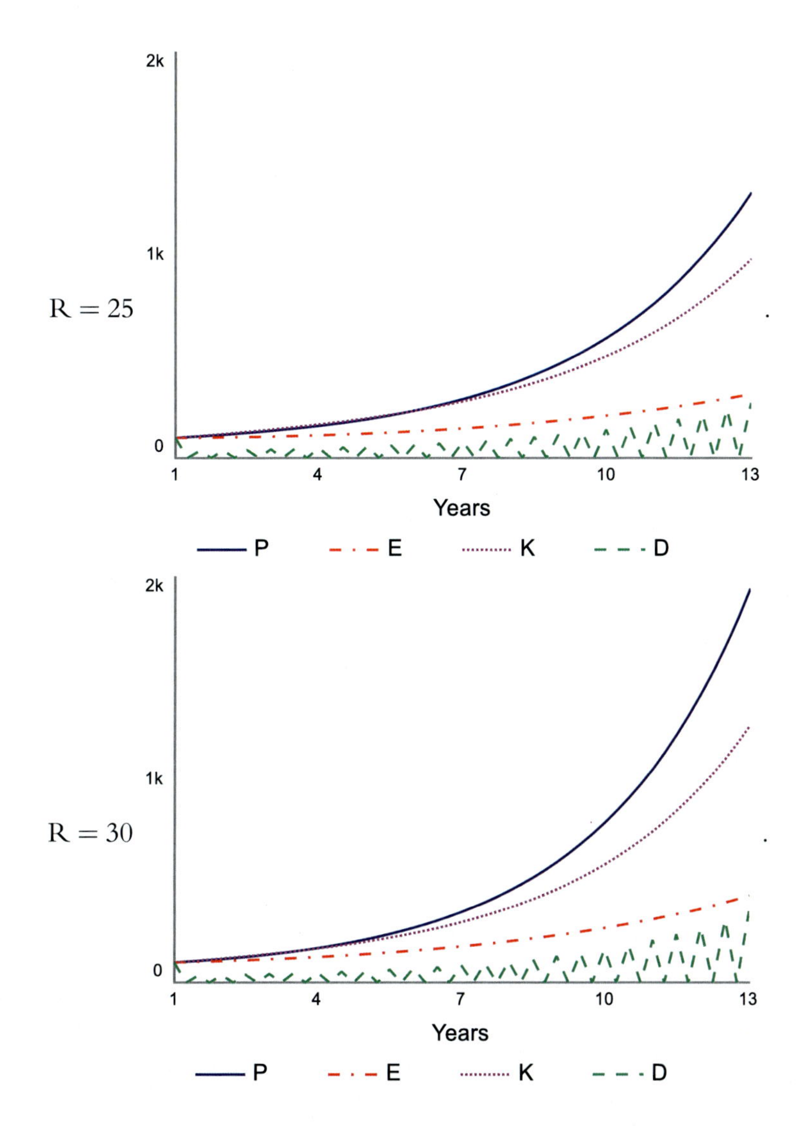

Low energy resources affect development at all levels of natural resource availability. Resource availability has some limits: (1) a limit on growth beyond which GDP starts growing, (2) a limit on efficiency below which GDP grows slower than the energy consumed at a low energy intensity. Additionally, as described later, there is a need for high energy resources and high natural resources to ensure development in which GDP has more grow than the energy consumed, thus leading to development with low energy intensity. One may notice that the limits of developmental behaviorial change are more complex in a space where both energy resources and natural resources vary.

Using data for a given economy, one may determine these limits and make decisions in accordance with the development measures.

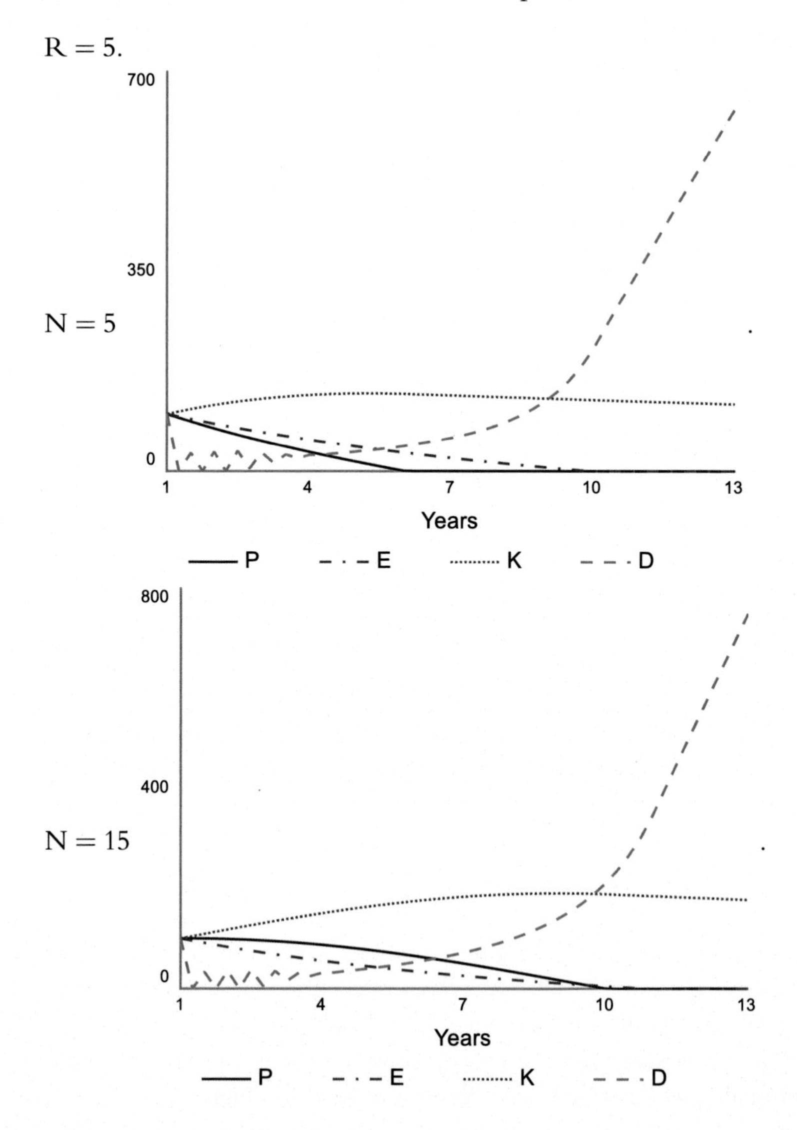

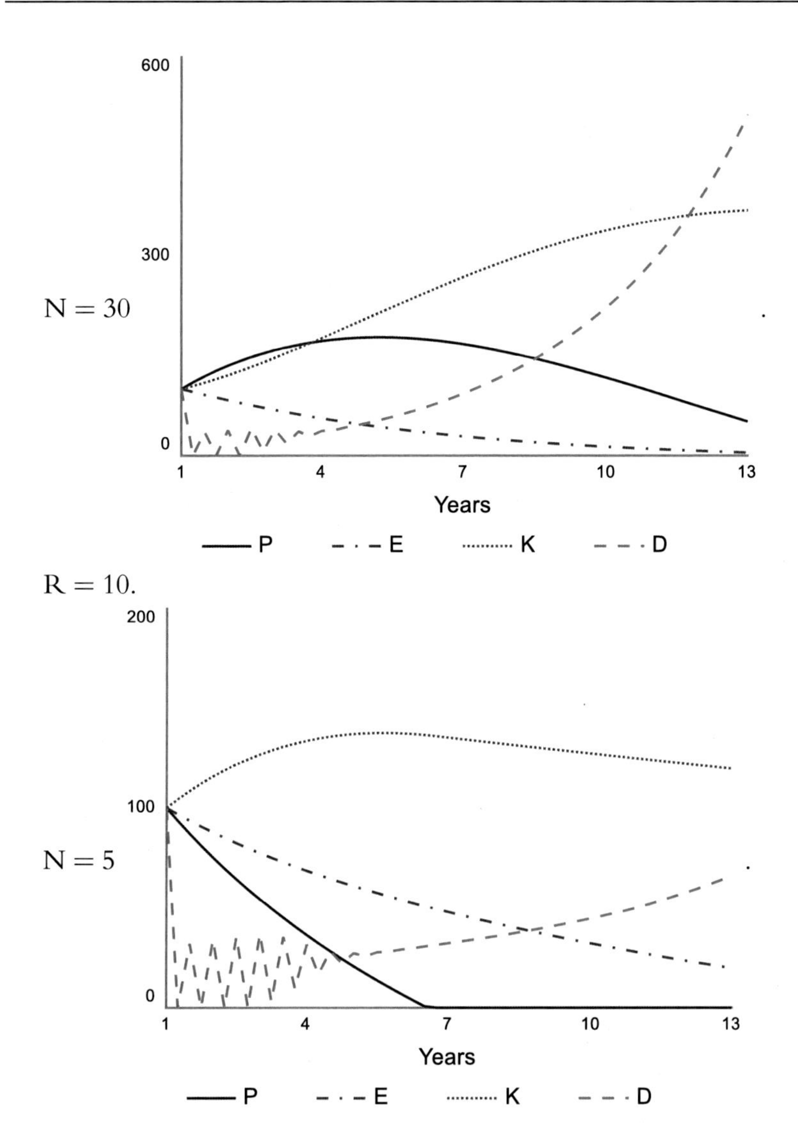
600
300
0
N = 30
1
4
7
10
13
Years
P
E
K
D
R = 10.
200
100
0
N = 5
1
4
7
10
13
Years
P
E
K
D

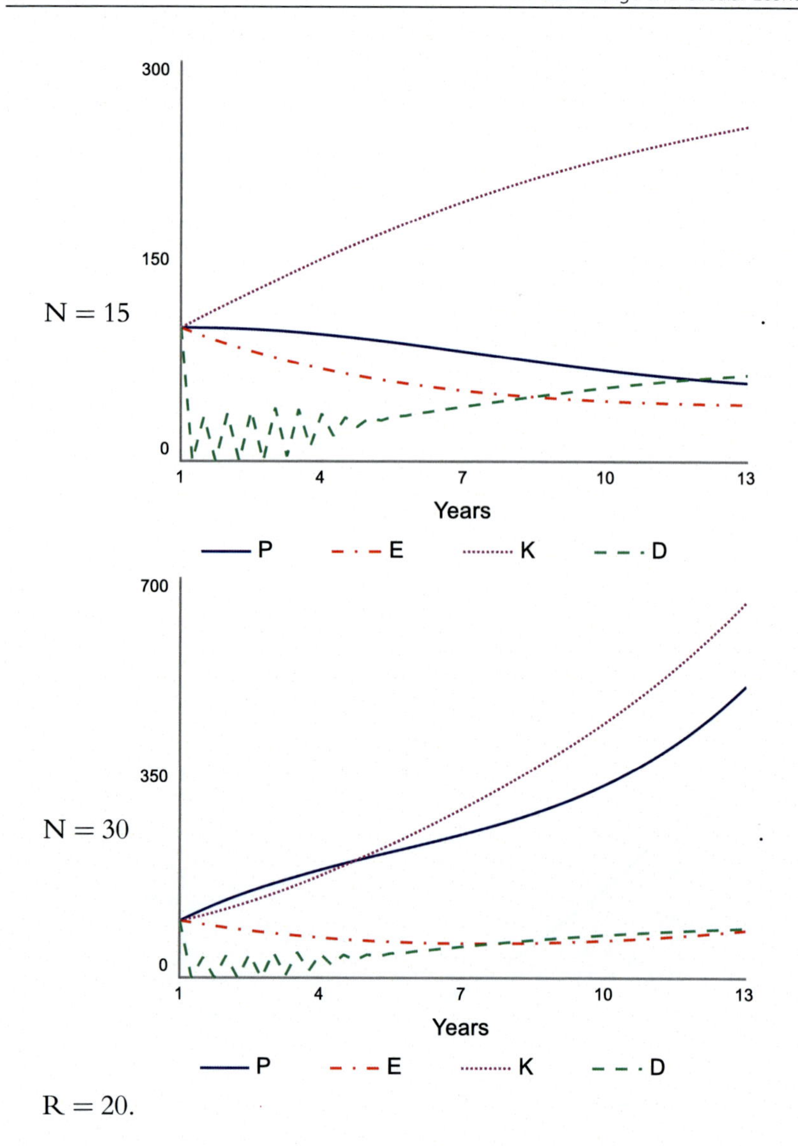
N = 15
300
150
0
1
4
7
10
13
Years
P
E
K
D
N = 30
700
350
0
1
4
7
10
13
Years
P
E
K
D
R = 20.

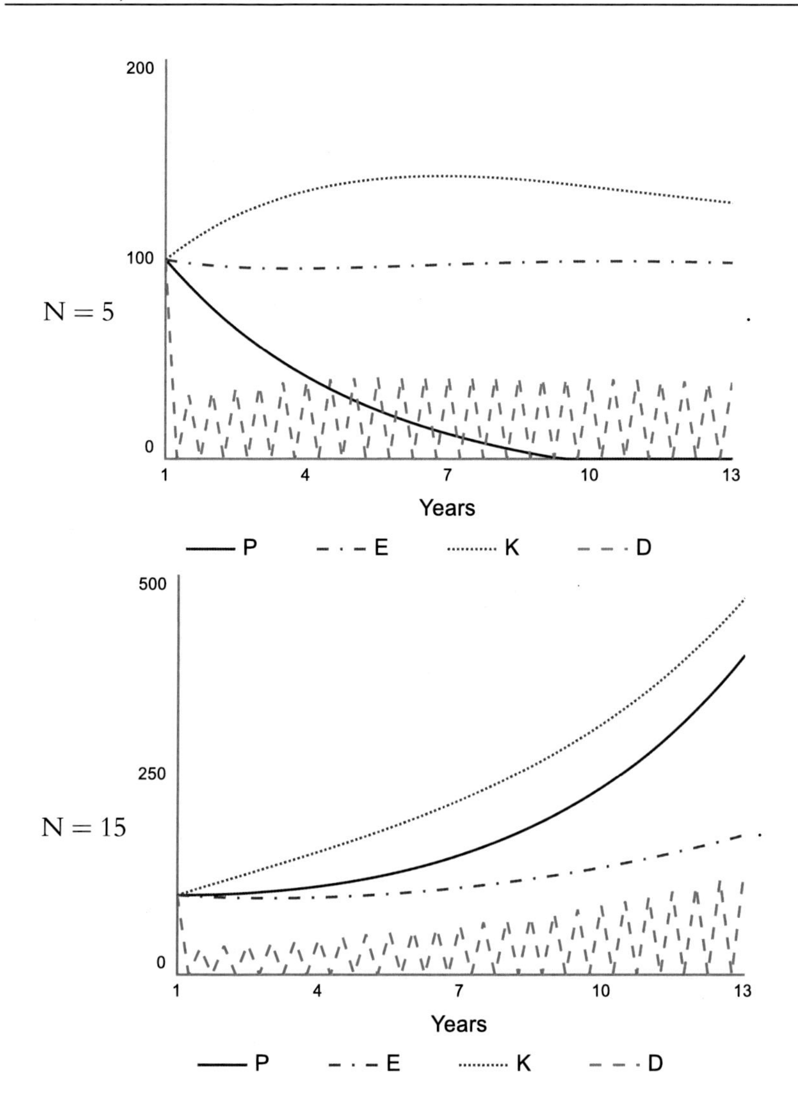
200
100
N = 5
0
1
4
7
10
13
Years
P
E
K
D
500
250
N = 15
0
1
4
7
10
13
Years
P
E
K
D

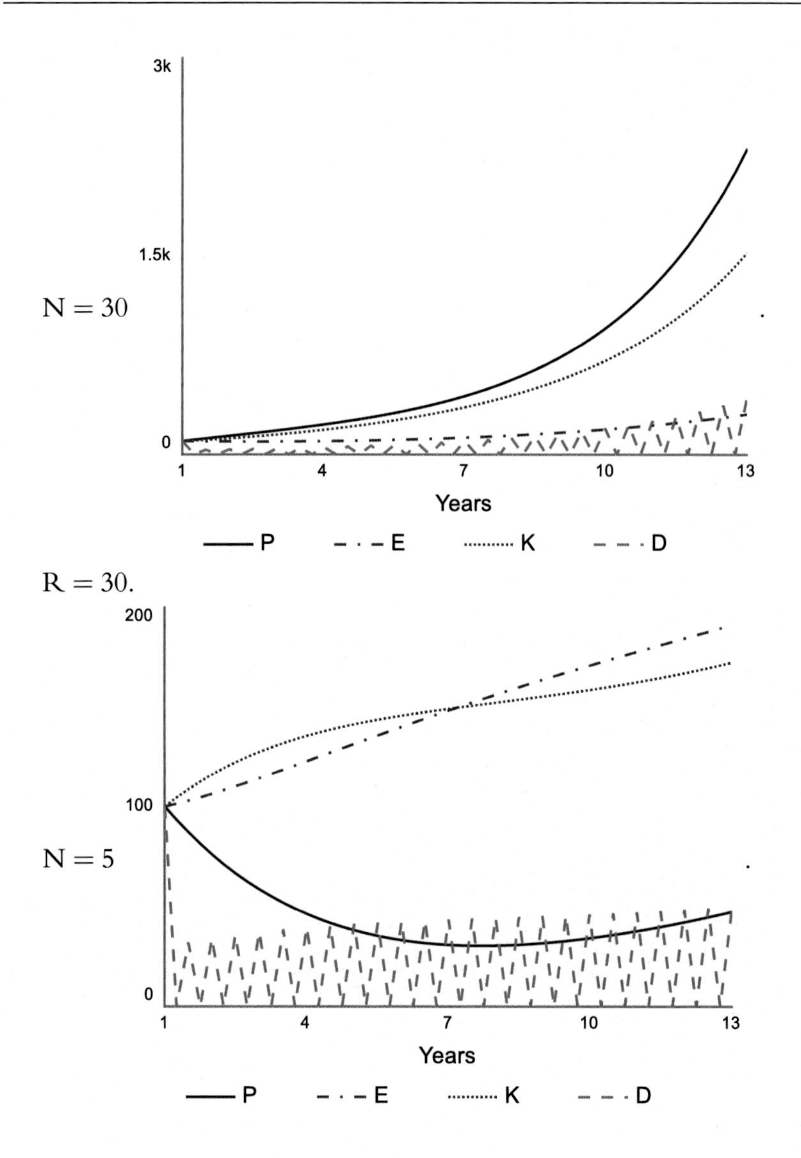
3k
1.5k
0
N = 30
1
4
7
10
13
Years
P
E
K
D
R = 30.
200
100
0
N = 5
1
4
7
10
13
Years
P
E
K
D

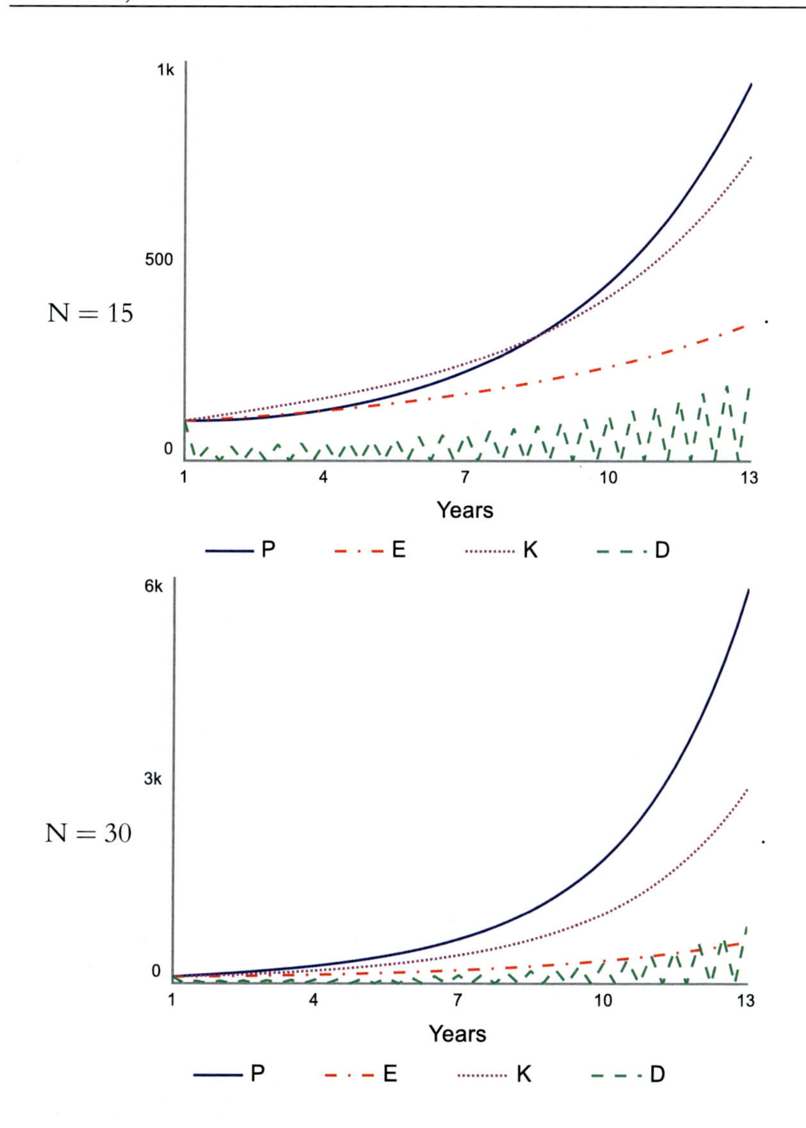

1.5.1 Use of capital

The long-term use of capital, i.e., its quality should be high (low replacement coefficient) to ensure growth, results from the figures below. That limits also exist here is shown in the figures as proof that the model is nonlinear and able not only to cover growth scenarios but also to determine behavior changes in development.

Creation of new capital has an innovation component that changes the importance of resources as well as the structure of the economy.

The inclusive society is further discussed below, and the impact on the environment is presented in the next chapter.

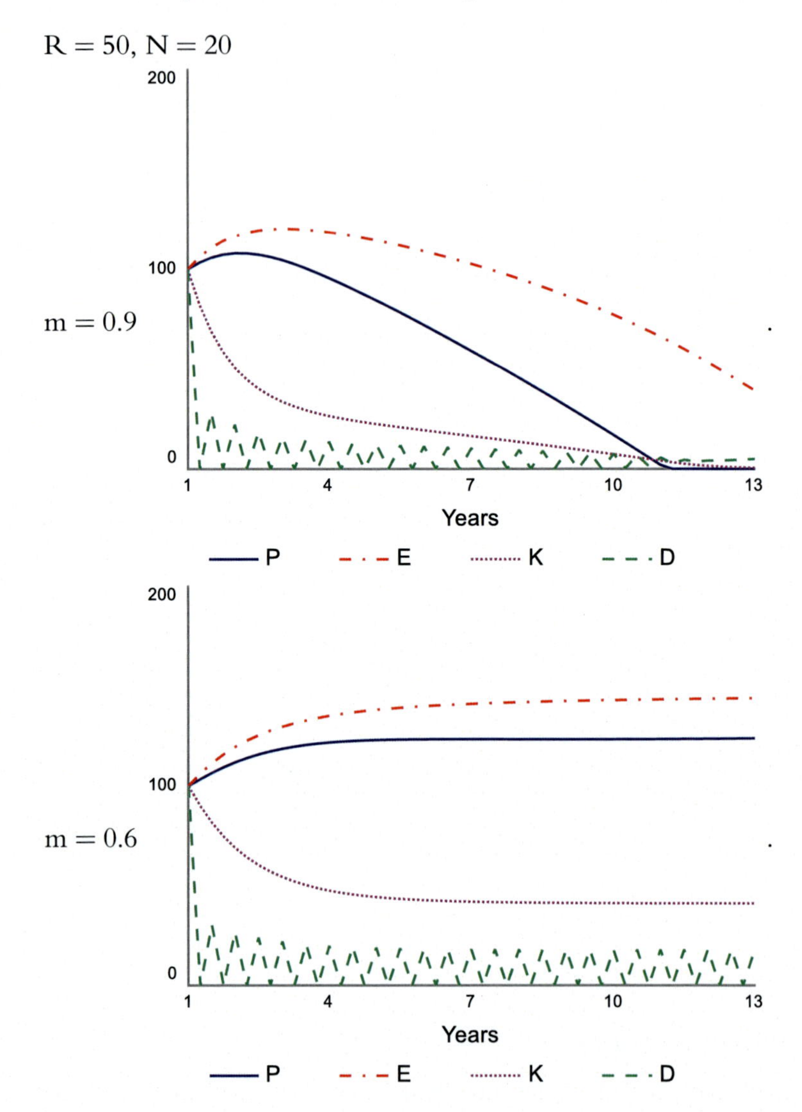

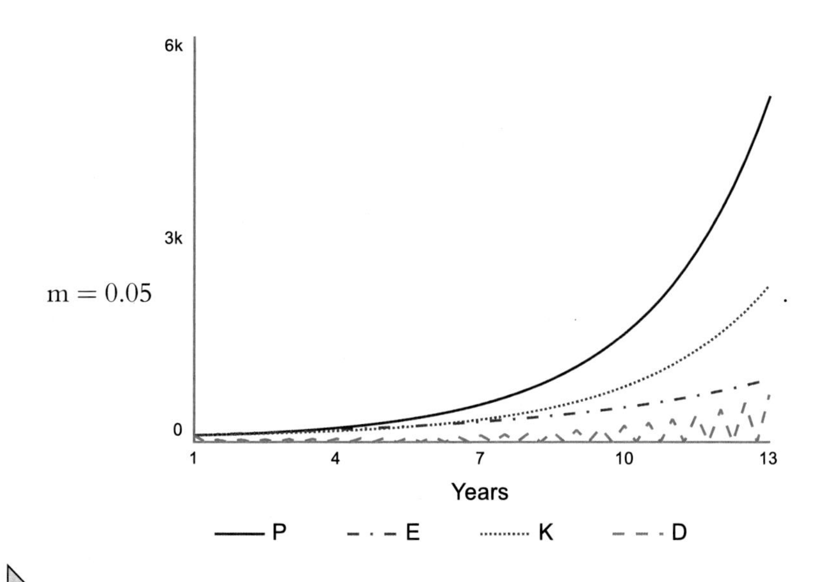

1.6 Is an inclusive society possible?

The second principle of thermodynamics, which by the way only specifies the increase in entropy in energy conversion processes, was much disputed when it came to the problem of processes in living and social systems. The third principle, discovered by Nernst, has been ignored until now. This principle specifies that entropy tends to zero when the temperature of the medium tends to absolute zero. As the temperature of the environment tends to absolute zero, the atoms of the body will be, at the minimum energy level, structured in a crystalline network characterized by an order well established by its symmetry. We could say that symmetry is maximum when entropy is zero. As you create the temperature of the system, atoms gain a higher average kinetic energy and their distance order begins to disappear, with the crystal passing into a liquid in which only local order is preserved. At an even higher temperature, we have the total disorder that is characteristic of gases. In all these processes, entropy, a measure of disorder, increases. If we consider that structures with life are not possible at the level of crystalline order or at the total disorder of gases but only at the level of the order of liquids, we will be surprised to find that they require a certain degree of local order concurrently with disorder at a distance, and the structures with local order fluctuate in space and time.

Biologically organized structures disappear if entropy increases above a certain value or if it decreases below a given value. What is important is maintaining the state with a certain degree of order, and for this, the exchange system with the environment, both substance and energy, causes entropy to grow to maintain a certain organization. So maintaining local order does not relate to entropy value but rather its growth rate. Local orders are maintained with a high entropy growth rate while others require a low entropy growth rate.

Prigogine, in a famous theorem, established that the rate of increase in entropy in chemical and molecular systems with life must be at the minimum compatible with the conditions imposed on the system.

Unlike living systems separated from the environment by a closed surface, human society coexists with the environment within a different topological structure because we do not have a clear separation of society and the environment but an interaction between social elements with the surrounding environment, as, for example, rational numbers (fractions) are embedded between real numbers. This slot does not prevent us from considering, as we know, human society as an open system that, in order to coexist with the environment, requires a certain entropy value, all the more so because it cannot exist without fulfilling the conditions for the existence of life.

But in its relationship with the environment, in order to maintain its organization, society causes the entropy of the environment to increase at a certain speed. However, this entropy growth rate depends on the internal structures of society, on how flows are reduced at the entries and exits of substances and energy between society and the environment. Therefore, cyclic closure of the processes is necessary to preserve the degree of the complex order of society.

We can characterize the evolution of human society from the stage of dissipative open system to the stage of closed system through entropy variation. This is created at high speed in the case of a dissipative society containing the production hypercycle. At the moment of the appearance of a new cycle, we have a rapid increase in entropy, after which the society, in a new stabilized structure, will evolve with entropy increasing at a lower rate.

If we were to characterize the state of society by an intensive parameter, let's say social progress, then each appearance of a new cycle becomes a bifurcation point where society can evolve in the sense of increasing progress, organize its structure, or decay through a rapid increase in entropy into a state of catastrophic disorganization.

The evolution to which we were led for the social infrastructure shows us that we must admit the tendency toward an antientropic society, in the sense that the speed of entropy growth is reduced with the occurrence of each new cycle allowed by the evolution of science. Without this, the degree of environmental pollution is reduced.

The term antientropic society, introduced by Valter Roman, acquires real content and value in this context.

We could even extract a law of evolution, with the following formulation:

"Society evolves toward an environmental degradation by incorporating a part of it in a cycle that is part of the system and leads to its growth, its order of organization, as well as to a decrease in the speed of entropy growth."

Human society, to escape the danger of pollution, has as its only solution the closing of itself within an open universe.

In the next chapter we will show that the effect on the environment from an increase in the organization of the economy may be assessed in terms of the environmental temperature increase produced by the evolution of various economies based on their energy intensity (or conversely, energy productivity) evolution.

Annex 1.1. Brief evolution of thermodynamics in economic systems

The representation of the economic process adopted by standard economics (see Samuelson and Nordhaus, 2010) will not be detailed here. To model the economic process as a thermodynamic system implies, therefore, the idea that the economy has complete reversibility that cannot possibly affect the environment of matter and energy in any way. The only feedback components of standard theory are those responsible for maintaining equilibrium, not irreversible changes. The obvious conclusion is that there is no need for bringing the environment into the classical representation of the reversible economic process.

The economic process is described as perpetual motion by Pigou in *The Economics of Stationary States* (1935): "In a stationary state, factors of production are stocks, unchanging in amount, out of which emerges a continuous flow, also unchanging in amount, of real income". Also, Georgescu-Roegen has the same idea, one of a constant flow characteristic for an unchanging structure embedded in Marx's simple reproduction model that also ignores the problem of how to obtain the primary input flows (Georgescu-Roegen, 1975).

In correlation emerges the view of Commoner, who was seriously concerned with ecological preservation in *The Closing Circle* (1974). Commoner proposed four ecological laws, the fourth of which is "there is no such thing as a free lunch": you cannot get something for nothing. This law also reminds us of the first law of thermodynamics, in that nothing can be obtained out of nothing. Thus, Commoner's law suggests a mechanical perspective envisioned by standard economics as well as the proponents of a circular economy, on that partially understands the irreversible nature of the ecological and economic processes of correlation.

By contrast, Georgescu-Roegen identifies the irreversible nature of the economic production process: "production represents a deficit in entropy terms: it increases total entropy by a greater amount than that which would result in the absence of any productive activity" (1971). Now Commoner's law must be replaced by Georgescu-Roegen's admonition: "you cannot get anything but at a far greater cost in low entropy."

Soddy (1961) also sees the essence of the problem: "This [energy] flow always occurs in one natural direction, and it can only be reversed in direction by making more energy flow downstream, so to speak, than flows up."

This trend toward recognizing the entropic nature of the economic process is a basic objective in the development of circular economy.

In 1945, Schrödinger added a note to Chapter VI, concluding "that we give off heat [thermal entropy] is not accidental, but essential. For this is precisely the manner in which we dispose of the surplus [thermal] entropy we continually produce in our physical life process" (Schrödinger 1967: 80). Schrödinger's deep insight shows that the disposal of surplus thermal entropy is necessary for living things to continue with life (Mayumi, 2001).

The school of nonequilibrium thermodynamics (Prigogine, 1961; Glansdorff and Prigogine, 1971; Nicolis and Prigogine, 1977) shows that the property of self-organization is specific to open dissipative systems—they must gather inputs from their environment and dispose of wastes into it.

A dissipative system has an organized structure of interaction controlled by two factors: (1) a dissipative structure—generating positive entropy flux to express its structures and functions; (2) an environment—providing a flux of low entropy energy and materials and changing high entropy energy and materials. Thus, cyclic operation is key to the survival of a dissipative system within a given environment. This property involves existential behavior by self-organizing systems, including dissipative ones—to survive they must irreversibly stress the environment in which they operate. For these complex systems to survive, they must learn and adapt to changes in environmental conditions. A socioecological system is a complex network of functional and structural components operating within a prescribed environment.

Sustainability requires the existence of a system for economic activities capable of closing the loops of primary inputs and outputs. This entails that recycling can be partially achieved if resources are available for it to be done and at increasing costs. The waste quantity can be reduced, but waste production is not avoidable, as with Georgescu-Roegen's description of deficit in entropy terms.

Moreover, that we live in a world with limits has been discussed by various economists, among them Kenneth Boulding, who said

"I am tempted to call the open economy the "cowboy economy," the cowboy being symbolic of the illimitable plains and, also associated with reckless, exploitative, romantic, and violent behavior, which is characteristic of open societies. The closed economy of the future might similarly be called the "spaceman" economy, in which the earth has become a single spaceship, without unlimited reservoirs of anything, either for extraction or for pollution, and in which, therefore, man must find his place in a cyclical ecological system which is capable of continuous reproduction of material form even though it cannot escape having inputs of energy.

Later, Boulding concludes:

"Anyone who believes exponential growth can go on forever in a finite world is either a madman or an economist."

References

Georgescu-Roegen, N., 1975. Energy and economic myths. Southern Economic Journal 41, 347–381.

Glansdorff, P., Prigogine, I., 1971. Thermodynamics Theory of Structure, Stability and Fluctuations. John Wiley and Sons, New York.

Mayumi, K., 2001. The Origins of Ecological Economics: The Bioeconomics of GeorgescuRoegen. Routledge, London.

Nicolis, G., Prigogine, I., 1977. Self-Organization in Nonequilibrium Systems. John Wiley and Sons, New York.

Prigogine, I., 1961. Introduction to Thermodynamics of Irreversible Processes, Second, Revised Edition. Interscience Publishers, New York.

Samuelson, P.A., Nordhous, W.D., 2010. Economics, nineteenth ed. MacGrawHill, New York.

Schrödinger, E., 1967. What is Life and Mind and Matter. Cambridge University Press, London.

Soddy, F., 1961. Wealth, Virtual Wealth, and Debt. Britons Publishing Company, London.

Further reading

Bigen, M., October 1971. Selforganization of Matter and the Evolution of Biological Macromolecules, Die Naturwissenschaften, ent.Lo.

Candelas, F., 1983. Ferticles production by the gravitational fields. In: Quantum Electrodinamics of Strong Fields, Ed. Greiner. Plenum Press.

Cherfas, J., 1991. Skeptics and visionaries examine energy saving. Science 251, 154–156.

Commoner, B., 1974. The Closing Circle. Bantam Books, New York.

Cullen, J.M., 2017. Circular economy: theoretical benchmark of perpetual motion machine? Journal of Industrial Ecology 21 (3), 483–486.
Georgescu-Roegen, N., 1971. The Entropy Law and the Economic Process. Harvard University Press, Cambridge MA.
Georgescu-Roegen, N., 1979. Energy analysis and economic valuation. Southern Economic Journal 44, 1023–1058.
Georgescu-Roegen, N., 1979. Laws of Entropy and Economic Progress, Col. Contemporary Ideas, Political Ed., Bucharest.
Giampietro, M., 1994. Using hierarchy theory to explore the concept of sustainable development. Futures 26 (6), 616–625.
Giampietro, M., 2003. Multi-Scale Integrated Analysis of Agro-Ecosystems. CRC Press, Boca Raton.
Giampietro, M., Mayumi, K., Sorman, A.H., 2012. The Metabolic Pattern of Societies: Where Economists Fall Short. Routledge, London.
Giorgi, H.M., April 1981. A Unified Theory of Elementary Particles and Forces. Scientific American, pp. 41–55.
Giorgi, M. and Glashow,S.L., Physical Review Letters, 32, 438, 1974.
Haas, W., Krausmann, F., Wiedenhofer, D., Heinz, M., 2015. How circular is the global economy? An assessment of material flows, waste production and recycling in the European union and the world in 2005. Journal of Industrial Ecology 19 (5), 765–777.
Herring, H., 1999. Does energy efficiency save energy? The debate and its consequences. Applied Energy 63, 209–226.
Jevons, W.S., 1865. The Coal Question (Reprint of the Third Edition-1906). New York.
Kelley, A.M., Khazzoom, J.D., 1987. Energy saving resulting from the adoption of more efficient appliances. Energy Journal 8 (4), 85–89.
Kirchherr, J., Reike, D., Hekkert, M., 2017. Conceptualizing the circular economy: an analysis of 114 definitions. Resources, Conservation and Recycling 127, 221–232.
Lomas, P.L., Giampietro, M., 2017. Environmental accounting for ecosystem conservation: linking societal and ecosystem metabolisms. Ecological Modelling 346, 10–19.
Mekonnen, M.M., Hoekstra, A.Y., 2014. Water footprint benchmarks for crop production: a first global assessment. Ecological Indicators 46, 214–223.
Mendoza, E. (Ed.), 1960. Reflections on the Motive Power of Fire by Sadi Carnot and Other Papers on the Second Law of Thermodynamics by É. Clapeyron and R. Clausius. Dover, New York.
Newman, P., 1991. Greenhouse, oil and cities. Futures 23 (3), 335–348.
Pigou, A.C., 1935. The Economics of Stationary States, London: Macmillan.
Polimeni, J., Mayumi, K., Giampietro, M., Alcott, B., 2007. The Jevons Paradox and the Myth of Resource Efficiency Improvements. The Earthscan, London.
Prigogine, I., 1978. Time, structure and fluctuations. Science 201 (4358), 777–785. Physics, Time and Becoming, Meson, Paris, 1982.
Purica, I., 1980. A model of the scientific-technical revolution within the mathematical theory of catastrophes. In: Revolutionary Froces in Science and Technology and the Development of Society, Ed. Politică, Bucharest.
Roman, V., 1978. Equilibrium and Imbalance, Scientific and Encyclopedic Ed., Bucharest.
Shi Itzu, H., Advances in Biophysics, 13, 195, 1979.
Smil, V., 2013. Making the Modern World. Wiley, New York.
Thok, R., 1975. Structural Stability and Morphogenesis. Benjamin Inc.
Tsuchida, A., 1982. Shigen Butsurigaku Nyumon (in Japanese) (An Introduction to Physics of Natural Resources). NHK Books, Tokyo.
Varela, F., 1984. Two principles for self-organization. In: In Self- Organization and Manarement of Social Systems. Springer Verlag.

CHAPTER TWO

Irreversible thermodynamics view of the need for a circular economy

The world we have created is a product of our thinking; it cannot be changed without changing our thinking.

Albert Einstein.

The second chapter introduces a method, based on the Gibbs equation in thermodynamics, of determining the environmental temperature increase induced by selected economies' evolution of energy intensity over long periods; the results are given in a table, while energy intensity is presented as a graph and digitized for calculations. The importance of resource recycling to temperature reduction is seen with this approach. Moreover, the rate of the temperature increase is also determined, opening the way to compare it with the time constant of environmental system recovery after an impact. From this point of view, the evolution of fast-developing emerging economies is seen in a new light relative to environmental resilience.

On a macroeconomic scale, the application of irreversible thermodynamics considerations shows that more efficient economic organization (i.e., local decrease in entropy) has an impact on the environment that can be measured by energy intensity dynamics. This component of environmental temperature increase may be estimated as the temperature increase in selected economies. Reductions in this temperature increase may be accomplished in this case through circular economic actions that make for more efficient exchanges of energy, resources, and waste with the environment and generate technologies that turn waste liabilities into assets. Moreover, the rate of temperature change is critical from the point of view of giving the environment and economies enough time to adapt.

2.1 Introduction

The relationship of human society with nature has been analyzed from various viewpoints for more than 3 decades since the first global agreement

Climate Change and Circular Economics
ISBN: 978-0-443-29969-8
https://doi.org/10.1016/B978-0-443-29969-8.00001-X

related to climate change. It is now clear that by exchanging energy and mass with the environment, human society is producing sizable effects in the last one. Besides energy and mass (call them material resources and waste), there is an effect from the exchange of entropy resulting from the increase in anthropic system organization, and hence, the local decrease in its entropy. On a product and technology scale, life cycle evaluations of the entropy source have been attempted as well as determinations of the value of the entropy production of a technological process or even an economy in correlation with energy or exergy (Bakshi (Ed), 2012). All these attempts recognize that better economic and technological organization effects a decrease in entropy that can be evaluated. At the scale of the whole economy of a given country, the impact can be measured in terms of temperature variations within an environment of increasing socioeconomic organization. This is driven by the evolution of commercial energy intensity in each economy.

On a long time scale, commercial energy intensity dynamics show a period of increase followed by one of decrease. The change is driven by the implementation of new technologies and by a more efficient economic structure that uses energy to create GDP. Thus in the increasing portion of energy intensity, the economy uses energy not only to ensure its survival as measured by GDP but also to change its organization into a more efficient one. Once this is done, the energy used produces more GDP than before (Purica, 1992). Based on this reasoning, the increasing part of the energy intensity evolution is proportional to the energy consumed from the environment, whereas the subsequent decrease is proportional to the level of the new, more efficient organization, i.e., to the lower entropy this achieves.

2.2 Irreversible thermodynamics approach

Considering the two systems—economy and environment—interconnected, one may write the irreversible thermodynamics equation to describe the interconnection (Guminski, 1964; deGroot, 1984). The terms describing the exchange of material resources are neglected for now:

$$dU/dt + dST/dt = 0$$

where dU is energy exchange, dS is entropy variation, T is temperature, and t is time.

This transforms into

$$T = (dU/dt)/(-dS/dt)$$

Now the increasing part of the energy intensity curve has a positive trend and is proportional to dU/dt, i.e., the extra energy used to create a more organized and hence efficient economy, and the decreasing part has a negative trend, i.e., the more efficient organization of the economy that created negative entropy in the open system, and is proportional to $-dS/dt$ (resulting in a positive dS/dt value). The preceding temperature formula results in a positive temperature value, i.e., an increase in environmental temperature (as an open system coupled with the economy) given by the economic evolution toward more efficient production of GDP, i.e., a more organized system that increases the entropy of the environment, which is seen as an increase in this one's temperature.

This effect of temperature increases in the environment due to more efficient organization of the economy must be compensated to avoid accumulation and eventually lead to the elimination of the local (subsystem) with low entropy.

The compensation may come from the term that we have willingly neglected above, i.e., the exchange of material resources between the economy and the environment. This adds one more term to the above equation:

$$dST/dt + dU/dt + \Sigma\kappa_{\iota}dm_i/dt = 0$$

where k_i is a constant of each resource i measured in energy per unit of resource, which describes the technologies available for using the said resource, with m_i as the quantity exchanged of resource i.

Analyzing this term leads to the consideration of resources versus waste treatment in the economic dynamic. The analysis is focused not on absolute values but on variations.

2.3 Circular economy

For example, to abate the temperature increase one must reduce the new denominator ($dU/dt + \Sigma\kappa_{\iota}dm_i/dt$). This is done by, e.g., reducing the velocity of transfer of primary resources by increasing product use time, i.e., $-dm/dt$. In economic terms this would mean, e.g., making more durable products with more possibilities to increase their use time by a better maintenance capability in their design and by better service in operation. In addition, waste recycling may reduce the rate of transfer of primary resources from/to the environment. Making steel from scrap metal rather than ore has an impact both on less primary resource transfer and on the smaller overall consumption of energy for metallurgical production. Additionally, the use of renewable energy conversion and high-energy-density fuels such as nuclear energy or fusion plants is likely to diminish the energy exchange rate with the environment.

From the above, a natural interconnection of the environment and the organization of the economy exists (in the sense of reducing entropy and increasing its capability to produce more GDP as a measure of development) in regard to the temperature increase in the environment and transfer of primary resources to the economic system. Moreover, new approaches based on nonequilibrium economics (Berger, 2009) and nonlinear decision models Purica (2010) are bringing better instruments to describe the process.

2.4 Estimating temperature increases and crises

To evaluate the order of magnitude of the temperature increase in the case of no resource transfer, data from the evolution of energy productivity curves for selected economies are used, as presented in Appendix 2.1. Energy productivity is the reverse of energy intensity, so the trend is first downward and then upward. Thinking in financial terms, this looks like an investment of energy into a more efficient economic organization that is paid back in energy once the new organizational structure is in place. Thus, in this case, the trend (slope) proportional to energy use is negative, and the one proportional to the entropy decrease is positive, resulting in the same positive increase in the temperature. Again, the energy return on investment is analyzed at a technological but not an economic scale.

The results are given in Table 2.1.

The estimated value of the temperature increase occurs over a period that represents, in the "energy investment for more efficient organization" interpretation, a payback time for the energy spent to create a better organization.

Table 2.1 Estimated temperature increases for selected economies (T in °C).

	dU	dS	$T = dU/-dS$
USA	−164.17	79.54	2.06
UK	−128.20	114.58	1.11
Germany	−64.10	102.27	0.62
Japan	−48.19	48.19	1.00
Italy	−21.97	40.81	0.53

Source: author's calculations based on data in Appendix 2.1.

For the selected countries, the estimated payback times are, in the UK: 160 years; the USA: 170 y; Germany: 140 y; Japan: 80 y; and Italy: 90 y.

Here it is important to introduce the speed of the temperature increase. This is an important parameter because it may be compared with the speed of environmental recovery after a given impact, resulting in the temperature increase associated with climate change. Looking at the values above, the speeds of temperature increases for the countries considered here are the following (in degrees celsius/century)—the UK: 0.69; the USA: 1.2; Germany: 0.44; Japan: 1.25; and Italy: 0.59.

One may notice that the shorter the payback time, the greater the speed. This suggests that new, fast-developing economies experience temperature changes at greater speed. If the environment has a recovery time (or resilience time constant) of 100 years and a temperature increase of 1°C/century and hence a speed of 1, the speed of the temperature increase induced by the new economies will not provide enough time for the environment to recover.

Along a different line, the trend of the temperature T( ) dynamic increases with more energy resource use and decreases with more natural resource recovery. We used the model in the previous chapter to assess these trends, as shown in Figs. 2.1 and 2.2. Note that the data in the figures only show trends not related to a specific economy. These trends confirm that resource recovery, renewable energy, and high-density energy resources, e.g., nuclear fission and fusion, are the solution for temperature reduction supporting the general concept of an inclusive society's economy. The concentration on emissions alone should be extended to an integrated approach to a fully circular economy.

Related to economic organization and temperature increases, one may review Figs. 2.3 and 2.4, where periods of increased temperatures occur during periods of decreasing energy intensity within large economies. Moreover, during the Second World War and its aftermath of economic reconstruction, temperature increases stagnated and then restarted after the middle of the 1970s (with the advent of new, more efficient economic activity) (see Fig. 2.5). A new wave of economies (emerging) is coming that may bring with it greater impacts on temperature increases if they do not learn from previous experience and if no common action is agreed to by everybody. The Paris COP21 (December 2015) agreement is a promising start, provided it will be applied.

One other situation where the temperature may decrease—judged by the evolution of CO_2 emissions—is economic crises (be they produced by

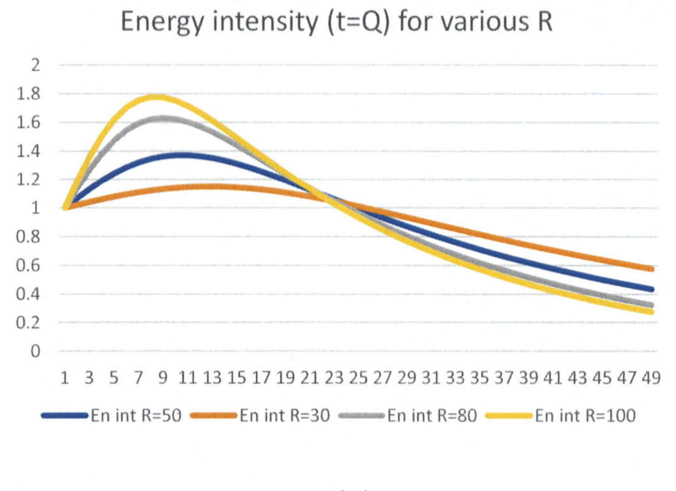

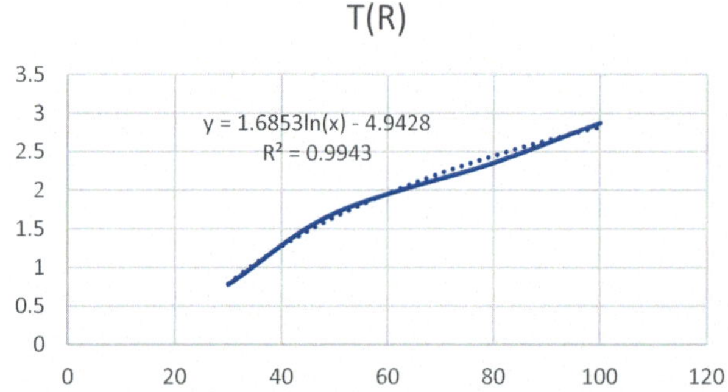

Figure 2.1 More classical energy resource use results in environmental temperature increases.

energy prices, financial evolution, pandemics, or other drivers). Figs. 2.3 and 2.6 are relevant in this sense.

Air traffic is representative of crisis periods. Usually, it represents about 3% of total emissions.

Total emission reductions during crisis periods, as presented in Fig. 2.6 above, match the air traffic represented in Fig. 2.3. It is normal to correlate reduced emissions with reduced temperatures. Thus, a reduction in the organization of the global economy affects decreases in environmental temperatures. The mechanism of GHG emissions may not be the only one contributing to this effect, but it is the most obvious in terms of the interconnection of cause and effect.

The occurrence in recent years of climate change that has generated conflicts and increasing risks of large confrontations that would disorganize

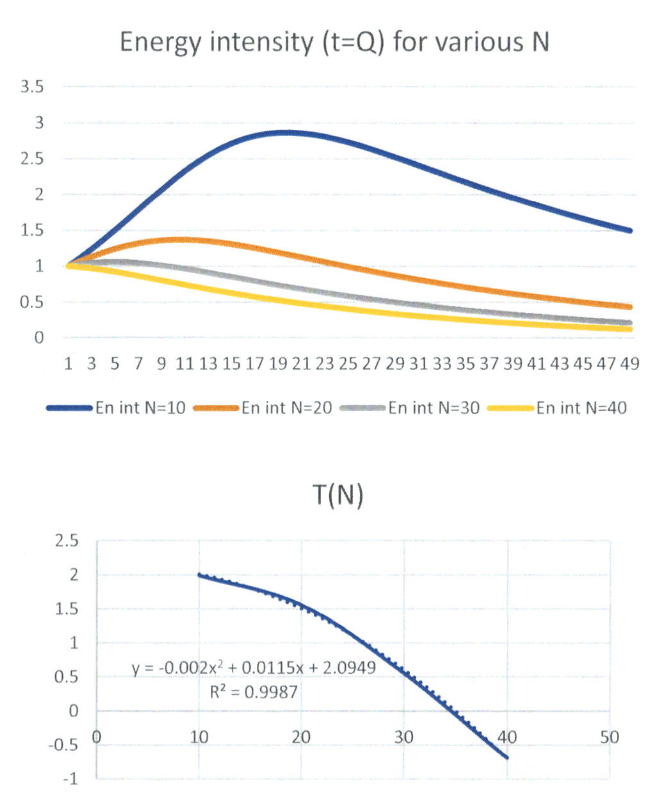

Figure 2.2 More recycling of resources results in environmental temperature decreases.

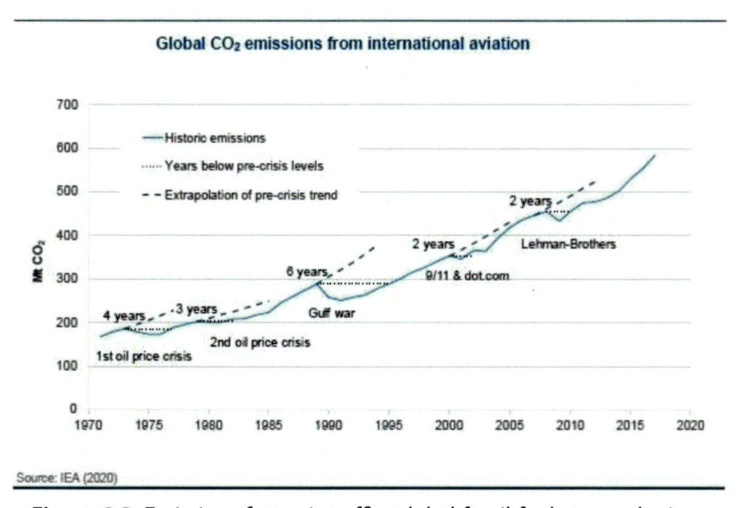

Figure 2.3 Emissions from air traffic, global fossil fuel use, and crises.

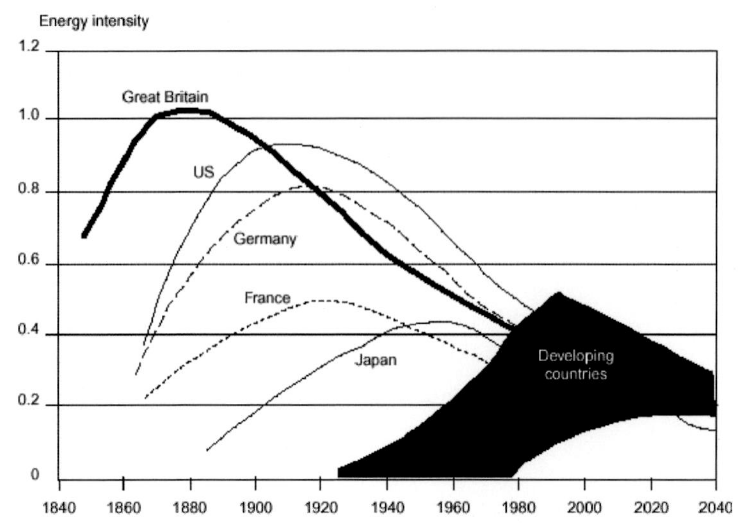

Figure 2.4 Energy intensity as reported to the British Parliament. *Source: Parliament Office of Science and Technology Notes, Climate Change, UK, 2007.*

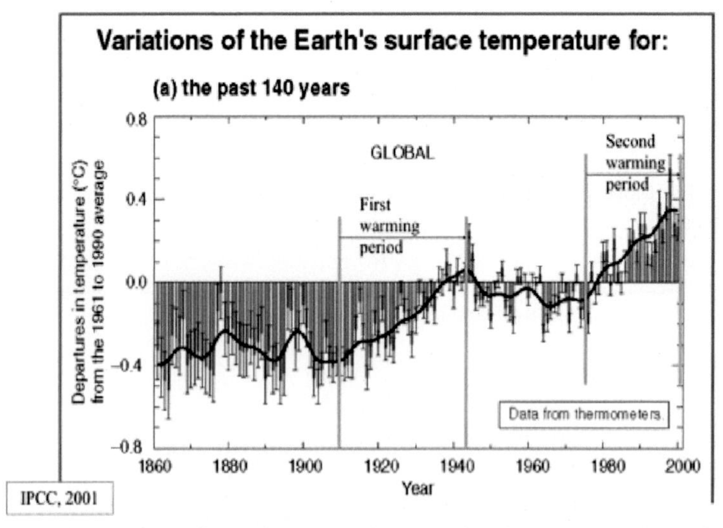

Figure 2.5 The Earth's surface temperature has risen by about 0.6 ± 0.2°C in the past century, with accelerated warming during the past 2 decades.

economic activity are, obviously, not a solution to generate more "anthropic" entropy that would reduce temperature increases.

The same picture, with Kanji symbols, was used to underline energy intensity and the late developer's advantage for the State Council Development Resource Center, China. The vertical axis is energy intensity (energy expended per unit of GDP).

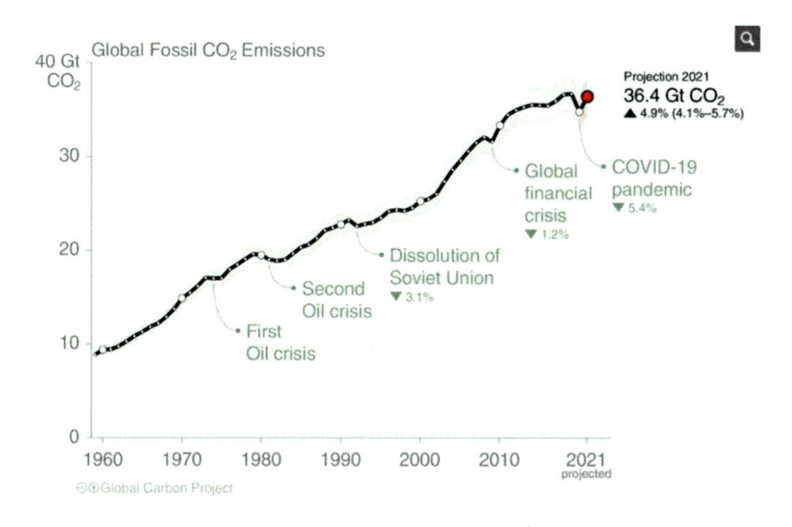

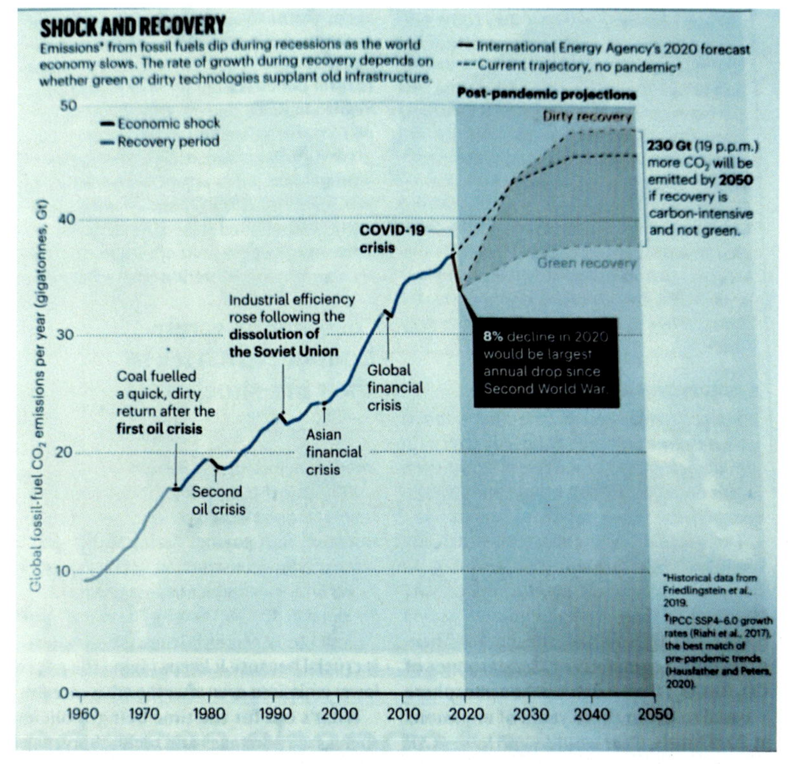

Figure 2.6 Emissions and crises.

2.5 Turning waste into assets—resource management policy and new technologies

The importance of resource recycling resulting from the above must be detailed and further analyzed for a better understanding of where we are today and, e.g., how to decide on the use of available resources, such as agricultural land for the partition between food for humans and biofuel for cars. Seeking resources is the basis of important human societal behavior, and decision-makers should consider the full complexity of the process where only emission reduction indicators may screen the supply chain and production impacts on the environment stemming from the use of resources.

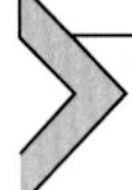

Appendix 2.1

See Figs. A.2.1.1 and A.2.1.2; Table A.2.1.1.

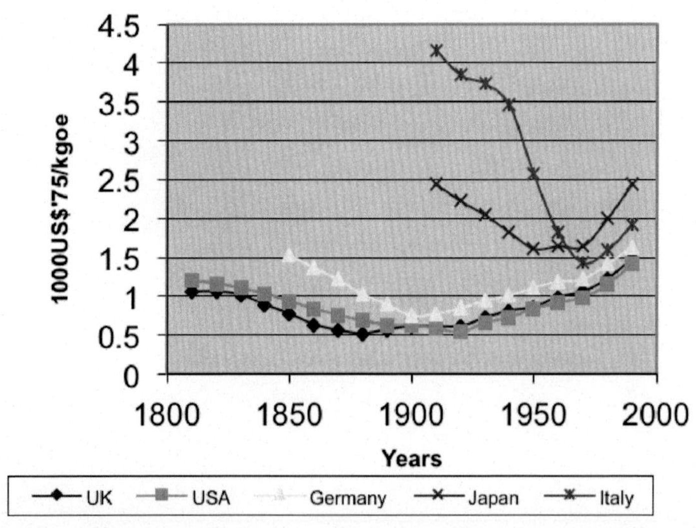

Figure A.2.1.1 Energy productivity of selected economies. *Source: Author's calculation based on Colombo, U., 2000. Energia. Storia e scenari, Universale Donzelli, Roma; Martin (1990).*

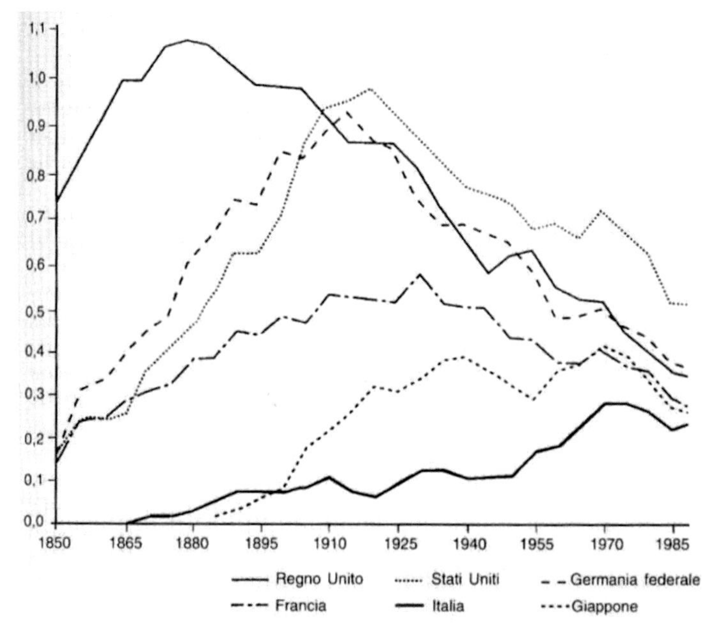

Figure A.2.1.2 Energy intensities of GDP, selected countries, 1850—1990 (tep/1980 dollars × 1000). *Source: Martin (1990); Toninelli, P.A., 2008. Energy Supply and Economic Development in Italy: The Role of the State-Owned Companies, Working Paper Series No. 146—October 2008, Dipartimento di Economia Politica Università degli Studi di Milano—Bicocca, http://dipeco.economia.unimib.it.*

Table A 2.1.1 Energy productivity of selected economies.

Year	UK	USA	Germany	Japan	Italy
1810	1.056	1.2			
1820	1.053	1.16			
1830	1.01	1.099			
1840	0.88	1.031			
1850	0.76	0.93	1.53		
1860	0.625	0.83	1.37		
1870	0.55	0.75	1.22		
1880	0.51	0.68	1.03		
1890	0.56	0.625	0.9		
1900	0.602	0.619	0.75		
1910	0.603	0.58	0.77	2.44	4.16
1920	0.604	0.53	0.84	2.22	3.84
1930	0.71	0.65	0.94	2.04	3.73
1940	0.8	0.72	1.01	1.82	3.45
1950	0.85	0.83	1.11	1.61	2.56
1960	0.98	0.91	1.19	1.64	1.81
1970	1.052	0.98	1.22	1.64	1.43
1980	1.22	1.15	1.41	2	1.59
1990	1.47	1.41	1.63	2.44	1.92

References

Bakchi, B.R., et al. (Eds.), 2012. Thermodynamics and the Destruction of Resources. Cambridge University Press, Cambridge, ISBN 9780521884556.

Berger, S. (Ed.), 2009. The Foundations of Non-equilibrium Economics. Routledge Advances in Heterodox Economics, New York, ISBN 9780415777803.

Colombo, U., 2000. Energia. Storia e scenari. Universale Donzelli, Roma.

de Groot, S.R., Mazur, P., 1984. Non-Equilibrium Thermodynamics. Dover Books on Physics, ISBN 0486647412.

Guminski, K., 1964. Termodinamica Proceselor Ireversibile, Editura Academiei, Bucuresti.

Parliament Office of Science and Technology Notes, Climate Change, UK, 2007.

Purica, I., 1992. Environmental change and the perception of energy system dynamics. In: ICTP-trieste, Conference on "Global Change and Environmental Considerations for Energy System Development", Proceedings.

Purica, I., 2010. Nonlinear Models for Economic Decision Processes. Imperial College Press, London, ISBN 9781848164277.

Toninelli, P.A., October 2008. Energy Supply and Economic Development in Italy: The Role of the State-Owned Companies, Working Paper Series No. 146. Dipartimento di Economia Politica Università degli Studi di Milano—Bicocca. http://dipeco.economia.unimib.it.

CHAPTER THREE

Resource materials and recycling technologies

In theory, theory and practice are the same. In practice, they are not.
Albert Einstein.

Effective recycling requires sophisticated knowledge about the components present in the end-of-life (EoL) products stream, which cannot be achieved with mixed recycling in broad categories. The European Academies' Science Advisory Council (EASAC) is satisfied that the case for moving toward a product-centric approach to collection and recycling is strong and should feature in the development of future EU policy (EASAC Priorities for critical materials for a circular economy, 2016).

- A more product-centric approach could improve currently low recovery levels by encouraging the building of EoL recovery into dedicated collection schemes to provide feedstock for specialized recycling. Options include deposit schemes, including return and recycling costs in the purchase price, trade-in which offers a financial reward for return, or contractual obligation. The current situation whereby much of Europe's e-waste leaves the EU (in many cases for informal and inefficient recycling in Asia or Africa) constitutes a significant leakage of critical metals requiring attention. A level playing field is needed so that low-quality recycling or avoiding recycling through legal loopholes does not continue to offer the cheapest option. The current proposals to amend the Extended Producer Responsibility requirements provide a mechanism to incorporate a special emphasis and priority on products containing economically significant quantities of critical materials. The supply of critical metals requires a baseline technological infrastructure that can recover metals from complex mixtures, thus extending the concept of criticality from that of individual elements to the infrastructure necessary for their cyclical use. The EU should evaluate the adequacy of the EU's "Critical Metallurgical Infrastructure" for the critical metals decided and consider measures to strengthen it.

Climate Change and Circular Economics
ISBN: 978-0-443-29969-8
https://doi.org/10.1016/B978-0-443-29969-8.00009-4

- Product design should consider the complexity of recycling and avoid incompatible metal mixtures or joints between product parts that hinder recycling. The EASAC notes, however, that trends driven by consumer convenience and demand continue to introduce burdens rather than facilitate recycling. The Commission should seek to engage consumer groups as well as manufacturers in a dialogue on ways to reduce or eliminate such inherent conflicts so that "design for resource efficiency" becomes standard practice.
- Developing effective recycling technology can require considerable investment. Particularly with critical materials, the circular economy policy must provide market signals that incentivize all companies to work toward a circular economy. The Horizon 2020 program should also support research and development on critical materials recovery and recycling ranging from the basic science underpinning the behavior of metals and their mixtures to novel separation and purification processes.

3.1 The main elements of the raw materials initiative

Pillar 1: Secure access to raw materials by ensuring undistorted world market conditions:

- Through diplomacy with resource-rich countries such as China and resource-dependent countries such as the United States and Japan for cooperation
- Through international cooperation via fora such as G8 and the Organisation for Economic Co-operation and Development (OECD) to raise awareness about the issues and create dialogue
- By making access to primary and secondary raw materials a priority for the EU trade and regulatory policy, to ensure that measures that distort open market trade such as restrictions of exports and dual pricing are eliminated

Pillar 2: Foster sustainable supply of raw materials from European countries by:

- Making sure the framework has the right conditions in place to prevent delays in permitting that can inhibit new projects

- Improving the European knowledge base for mineral deposits. Long-term access to these deposits should be considered during land use planning
- Better exchanging information between countries through networking among the national geological surveys
- Promoting research projects with a focus on extraction and processing (7th Framework Programme, continued in Horizon 2020) and making funding available for projects
- Increasing the number of skilled personnel through cooperation with universities and increasing public awareness of the importance of domestic materials.

Pillar 3: Reduce the EU's consumption of primary raw materials by:

- Improving resource efficiency, such as improved product design through the Eco-Design Directive
- Decreasing the volume of materials lost through illegal exporting to secure secondary raw materials, which will also require solid relations with third-countries to ensure the enforcement of waste shipment regulations
- Increasing reuse and recycling through legislation, standards and labeling, financing, knowledge sharing, and other means. The three tables that follow are providing a synthesis of the recycling situation/potential in the EU.

Recycling rates in EU-20 CRM LIST.

Recycling rate (according to UNEP)	**Critical materials on EU list**
<1%	Beryllium, gallium, germanium, indium, osmium, rare earths
1%–10%	Antimony
>10%–25%	Ruthenium, tungsten
>25%–50%	Magnesium, iridium
>50%	Cobalt, niobium, platinum, palladium, rhodium, chromium

Courtesy: Wellmer, F., Hagelüken, C., 2015. The feedback control cycle of mineral supply, increase of raw material efficiency, and sustainable development. Minerals 5, 815–836.

Potential for extending the lifetime of supply through recycling.

Element	Business-as-usual burn-off time[a] (years from 2011)	50% recycled (years from 2011)	70% recycled (years from 2011)
Nickel	42	b	209
Copper	31	b	157
Zinc	20	37	61
Manganese	29	46	229
Indium	19	38	190
Lithium	25	49	245
Tin	20	30	150
Molybdenum	48	72	358
Lead	23	23	90
Niobium	45	72	360
Helium	9	17	87
Arsenic	31	62	309
Antimony	25	35	175
Gold	48	48	71
Silver	14	b	43
Rhodium	44	b	132

[a]Burn-off time is defined as the estimated extractable resources divided by the present net extraction rate.
[b]The current recycling rate is already 50% or above.

Production rates, recoverable amounts, recycling rates, and years remaining in supply in current reserves.

Metal	Global production 2012 (tonnes per year)	Recoverable reserves (tonnes)	Recycling rate (%)	Reserve-to-production ratio (years)
Iron	1,400,000,000	340,000,000,000	60	242
Aluminum	44,000,000	22,400,000,000	75	436
Manganese	18,000,000	1,030,000,000	45	57
Chromium	16,000,000	437,000,000	22	27
Copper	16,000,000	558,000,000	60	35
Zinc	11,000,000	1,110,000,000	20	101
Lead	4,000,000	693,000,000	65	173
Nickel	1,700,000	96,000,000	60	56
Titanium	1,500,000	600,000,000	20	400
Zirconium	900,000	60,000,000	10	67
Magnesium	750,000	200,000,000,000	40	260,000
Strontium	400,000	1,000,000,000	0	2500

Production rates, recoverable amounts, recycling rates, and years remaining in supply in current reserves.—cont'd

Metal	Global production 2012 (tonnes per year)	Recoverable reserves (tonnes)	Recycling rate (%)	Reserve-to-production ratio (years)
Tin	300,000	76,200,000	20	254
Molybdenum	280,000	22,500,000	40	80
Vanadium	260,000	19,400,000	40	75
Lithium	200,000	40,000,000	10	200
Antimony	180,000	7,000,000	15	39
Rare earths	130,000	100,000,000	15	770
Cobalt	110,000	11,600,000	40	105
Tungsten	90,000	2,900,000	40	32
Niobium	68,000	3,972,000	60	58
Silver	23,000	1,308,000	80	57
Yttrium	8900	540,000	10	61
Bismuth	7000	360,000	15	51
Gold	2600	135,000	95	52
Selenium	2200	171,000	0	78
Cesium	900	200,000,000	0	220,000
Indium	670	47,100	40	70
Tantalum	600	58,500	25	97
Gallium	280	5200	15	19
Beryllium	250	80,000	20	320
Palladium	200	36,000	60	180
Platinum	180	44,100	70	245
Germanium	150	12,500	30	83
Tellurium	120	11,080	0	92
Rhenium	55	4190	85	84
Rubidium	22	5,000,000	0	227,000
Thallium	10	380,000	0	38,000

From Sverdrup, H., Ragnarsdottir, K., 2014. Natural resources in a planetary perspective. Geochemical Perspectives 3, 129–336.

10 recoverable reserves will of course change, so such estimates are only indicative of trends based on data available at the time of the study.

3.2 Improving recycling rates

A fundamental strategy in securing the supply of critical materials is ensuring that their use is as efficient and cyclical as possible. The amount of some critical metals present in consumer goods sold during a year is significant (quantities can range from 4% to 20% of the annual mine production of the metal (Hagelüken, 2014)). However, this is distributed across a wide range of consumer products and a precondition for recycling is for sufficient quantities of these dispersed products to be collected. There is also a lag between the input of the material into the consumer goods "technosphere" and the availability of recycling. For instance, the use of PGM in automotive catalysts started in the mid-1990s and as a result of the relatively long lifetime of cars, 1100 tons of PGM are in use within Europe (Hagelüken, 2014) and amounts available for recycling will be increasing as post-catalyst era cars are scrapped. In the absence of an effective recycling system (for instance due to lack of removal from all cars before shredding), these valuable elements will be lost. Other sources of high demand for critical materials relate to renewable energy technologies (e.g., wind generators and solar panels) and will remain in operation for many years before they are available for recycling in large quantities. Recycling infrastructure must anticipate these future trends.

Despite this lag in the supply of EoL products from some sectors, substantial amounts of smaller and shorter-lived applications using critical materials already enter the waste stream yet have very low rates of circularity.

Quantities are already strategically and economically important if appropriate recycling methods and technologies can be applied. For instance, Du and Graedel (2011) estimate globally that in 2007 between 1000 and 3000 tons of praseodymium, neodymium, and yttrium went to landfill. Some countries are focusing attention on specific elements and their material flows: for instance, Japan introduced a rare earths strategy in 2009.

3.3 Japan and rare earths in permanent magnets

Faced with supply risks triggered by Chinese restrictions on exports of rare earths, Japan introduced a "strategy for ensuring stable supplies of rare metals" in 2009. Enhancing recycling was one of the four pillars with a budget allocated for research into recycling technologies for rare earths (via NEDO). One of the most important uses is permanent magnets. Methods for recycling rare earth magnets are long established (although they have only recently become cost-effective), using a process involving molten magnesium to extract the rare earths. The recycled metals are suitable for the manufacture of new magnets with only a small degradation in performance compared with new magnets. Organizations supported by research funding (e.g., Japan Rare Earths) have also developed new methods for the extraction of rare earths from recyclable magnets using proprietary processes that are simpler than previous methods and allow the rare earths recovered to retain more of their functional properties. Hitachi has developed equipment that greatly improves the efficiency of extracting magnets. Toyota also received the Prime Minister's prize for achievements in 3R (reduce, reuse, and recycle) for its technology which enables rare metals in hybrid vehicles to be recovered for use in new motor magnets. Mitsubishi Materials has developed new technology to recycle rare earth magnets from compressors in air conditioners and washing machine motors. Japan Metals and Chemicals Company and Honda have developed a recycling facility capable of recovering rare earths from batteries. Such government initiatives offer options for improving the recycling rate and avoiding scrap materials ending up in generic scrap metal waste streams.

3.4 Managing resources

A more product-centric approach could improve on current low levels of recovery by encouraging recovery to be built into dedicated collection schemes providing feedstock for specialized recycling. Options include deposit schemes, including return and recycling costs in the purchase price, trade-in which offers a financial reward for return, or contractual obligation.

The current situation whereby much of Europe's e-waste leaves the EU (in many cases for informal and inefficient recycling in Asia or Africa) constitutes a significant leakage of critical metals requiring attention. A level playing field is needed so that low-quality recycling or avoiding recycling through legal loopholes does not continue to offer the cheapest option. For particularly critical materials, the viability of labeling schemes to trace metals from the mine to the market should be evaluated. The current proposals to amend the Extended Producer Responsibility requirements provide a mechanism to incorporate special emphasis and priority on products containing economically significant quantities of critical materials.

Supply of critical metals requires a baseline technology infrastructure that can recover metals from complex mixtures, thus extending the concept of criticality from that of the individual elements to the infrastructure necessary for their cyclical use.

The US economy's dependence on critical minerals is depicted in the table below.

Here too the strive to find these minerals goes beyond the planetary limit, with the recent developments in space technologies, e.g., mining metallic asteroids or seeking isotopes and minerals on the moon.

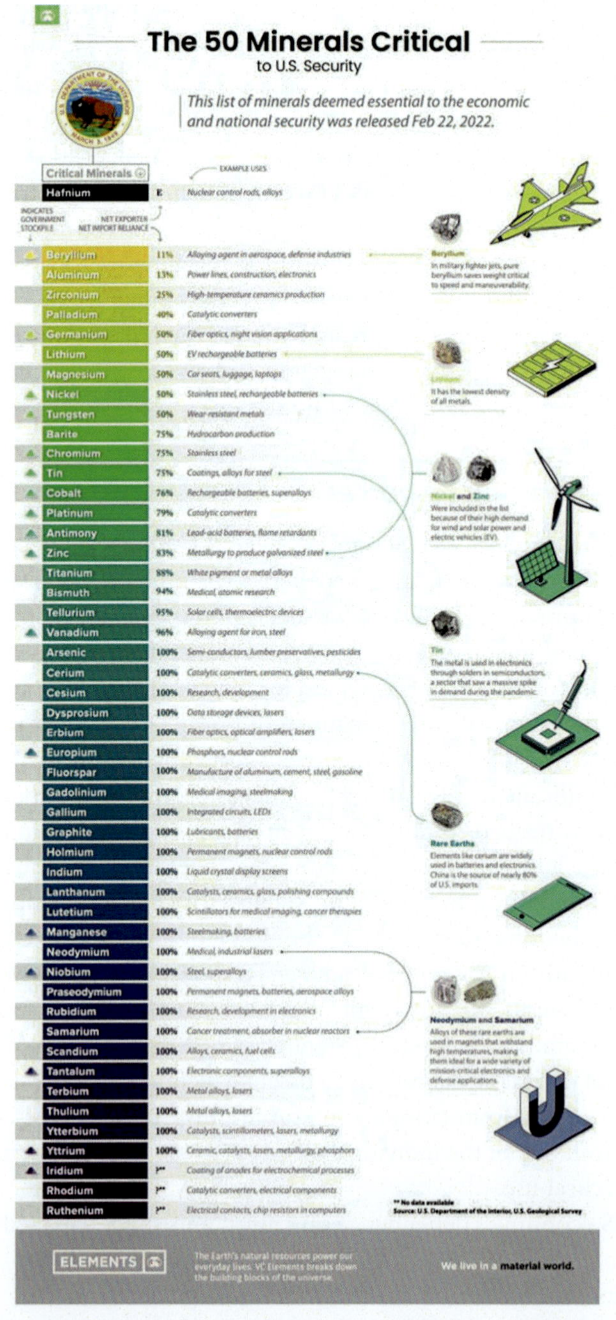

Let's see below where some of those minerals are coming from. Published on August 4, 2023 by Niccolo Conte, Graphics/Design: Pernia Jamshed, Technology Metals.

3.4.1 Charting America's import reliance on key minerals

The push toward a more sustainable future requires various key minerals to build the infrastructure of the green economy. However, the United States relies heavily on nonfuel mineral imports causing potential vulnerabilities in the nation's supply chains.

Specifically, the United States is 100% reliant on imports for at least 12 key minerals deemed critical by the government, with China being the primary import source for many of these along with many other critical minerals.

This graphic uses data from the US Geological Survey (USGS) to visualize America's import dependence for 30 different key nonfuel minerals along with the nation that the United States primarily imports each mineral from.

3.4.2 US import reliance by mineral

While the United States mines and processes a significant amount of minerals domestically, in 2022 imports still accounted for more than half of the country's consumption of 51 nonfuel minerals. The USGS calculates a net import reliance as a percentage of apparent consumption, showing how much of US demand for each mineral is met through imports.

Of the most important minerals deemed by the USGS, the United States was 95% or more reliant on imports for 13 different minerals, with China being the primary import source for more than half of these.

Mineral	Net import reliance as percentage of consumption	Primary import source (2018–21)
Arsenic	100%	CN China
Fluorspar	100%	MX Mexico
Gallium	100%	CN China
Graphite (natural)	100%	CN China
Indium	100%	KR Republic of Korea
Manganese	100%	GA Gabon
Niobium	100%	BR Brazil
Scandium	100%	EU Europe
Tantalum	100%	CN China
Yttrium	100%	CN China
Bismuth	96%	CN China
Rare earths (compounds and metals)	95%	CN China
Titanium (metal)	95%	JP Japan
Antimony	83%	CN China
Chromium	83%	ZA South Africa
Tin	77%	PE Peru
Cobalt	76%	NO Norway
Zinc	76%	CA Canada
Aluminum (bauxite)	75%	JM Jamaica
Barite	75%	CN China
Tellurium	75%	CA Canada
Platinum	66%	ZA South Africa
Nickel	56%	CA Canada
Vanadium	54%	CA Canada
Germanium	50%	CN China
Magnesium	50%	IL Israel
Tungsten	50%	CN China
Zirconium	50%	ZA South Africa
Palladium	26%	RU Russia
Lithium	25%	AR Argentina

These include rare earths (a group of 17 nearly indistinguishable heavy metals with similar properties) which are essential in technology, high-powered magnets, electronics, and industry, along with natural graphite which is found in lithium-ion batteries.

These are all on the US government's critical mineral list which has a total of 50 minerals, and the United States is 50% or more import reliant for 43 of these minerals.

Some other minerals on the official list which the United States is 100% reliant on imports for are arsenic, fluorspar, indium, manganese, niobium, and tantalum, which are used in a variety of applications like the production of alloys and semiconductors along with the manufacturing of electronic components like LCD screens and capacitors.

3.5 China's gallium and germanium restrictions

America's dependence on imports for various minerals has resulted in a new challenge resulting from China's announced export restrictions on gallium and germanium that took effect on August 1, 2023. The United States is 100% import-dependent on gallium and 50% import-dependent on germanium.

These restrictions are seen as a retaliation against US and EU sanctions on China which have restricted the export of chips and chip-making equipment.

Both gallium and germanium are used in the production of transistors and semiconductors along with solar panels and cells, and these export restrictions present an additional hurdle for critical US supply chains of various technologies that include LED lights and fiber-optic systems used for high-speed data transmission.

The restrictions also affect the European Union, which imports 71% of its gallium and 45% of its germanium from China. It is another stark reminder to the world of China's dominance in the production and processing of many key minerals.

The announcement of these restrictions has only highlighted the importance for the United States and other nations to reduce import dependence and diversify supply chains of key minerals and technologies.

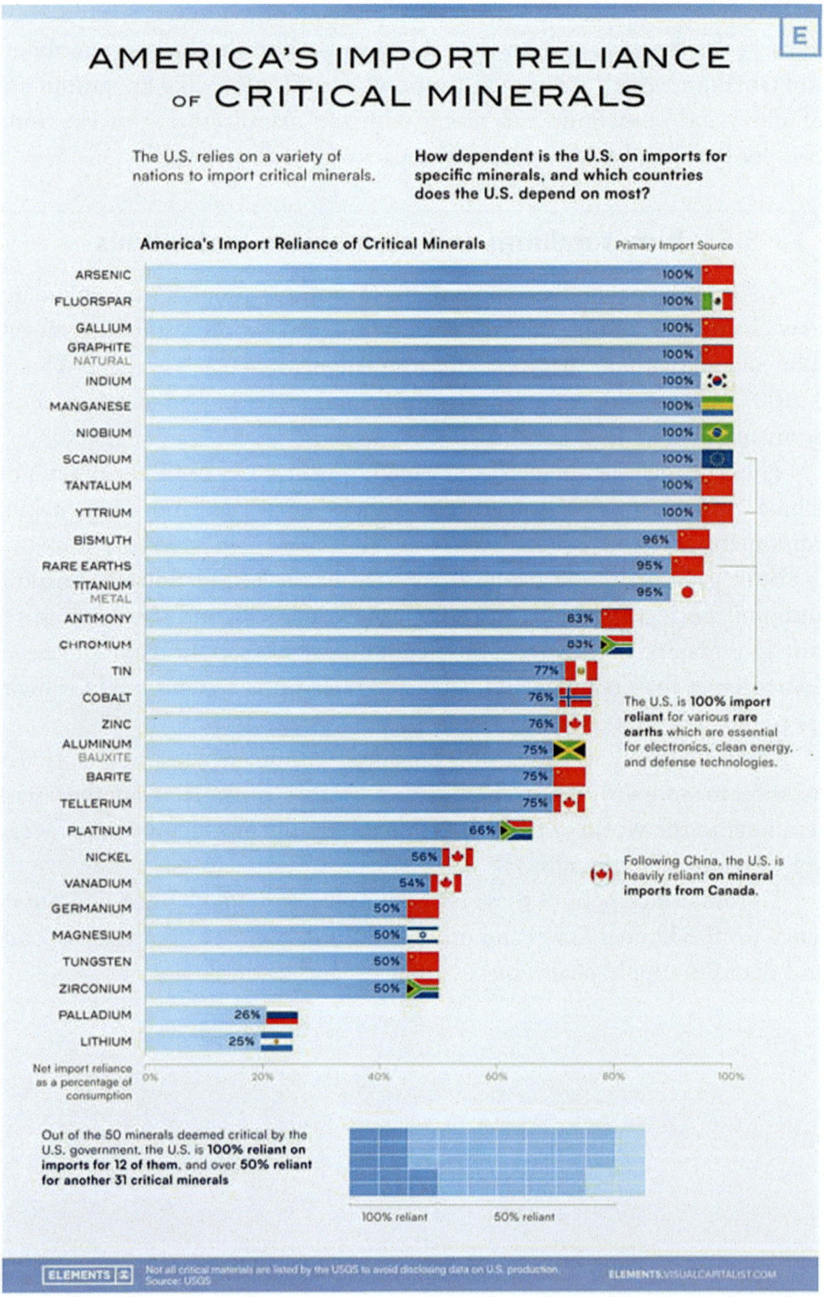

E

AMERICA'S IMPORT RELIANCE OF CRITICAL MINERALS

The U.S. relies on a variety of nations to import critical minerals.

How dependent is the U.S. on imports for specific minerals, and which countries does the U.S. depend on most?

America's Import Reliance of Critical Minerals

Mineral	Net import reliance as a percentage of consumption
ARSENIC	100%
FLUORSPAR	100%
GALLIUM	100%
GRAPHITE NATURAL	100%
INDIUM	100%
MANGANESE	100%
NIOBIUM	100%
SCANDIUM	100%
TANTALUM	100%
YTTRIUM	100%
BISMUTH	96%
RARE EARTHS	95%
TITANIUM METAL	95%
ANTIMONY	83%
CHROMIUM	83%
TIN	77%
COBALT	76%
ZINC	76%
ALUMINUM BAUXITE	75%
BARITE	75%
TELLERIUM	75%
PLATINUM	66%
NICKEL	56%
VANADIUM	54%
GERMANIUM	50%
MAGNESIUM	50%
TUNGSTEN	50%
ZIRCONIUM	50%
PALLADIUM	26%
LITHIUM	25%

Primary Import Source

Net import reliance as a percentage of consumption: 0% 20% 40% 60% 80% 100%

The U.S. is **100% import reliant** for various **rare earths** which are essential for electronics, clean energy, and defense technologies.

Following China, the U.S. is heavily reliant **on mineral imports from Canada.**

Out of the 50 minerals deemed critical by the U.S. government, the U.S. is **100% reliant on imports for 12 of them**, and over **50% reliant for another 31 critical minerals**

100% reliant 50% reliant

ELEMENTS Not all critical materials are listed by the USGS to avoid disclosing data on U.S. production.
Source: USGS

ELEMENTS.VISUALCAPITALIST.COM

3.6 Rebirth of nuclear and the needed resources

On a different line, the reduction of emissions is producing a rebirth of nuclear power technologies. This is likely to increase the use of Uranium resources as well as the thorium ones. The evolution of uranium for nuclear power fuel is presented in the figure below.

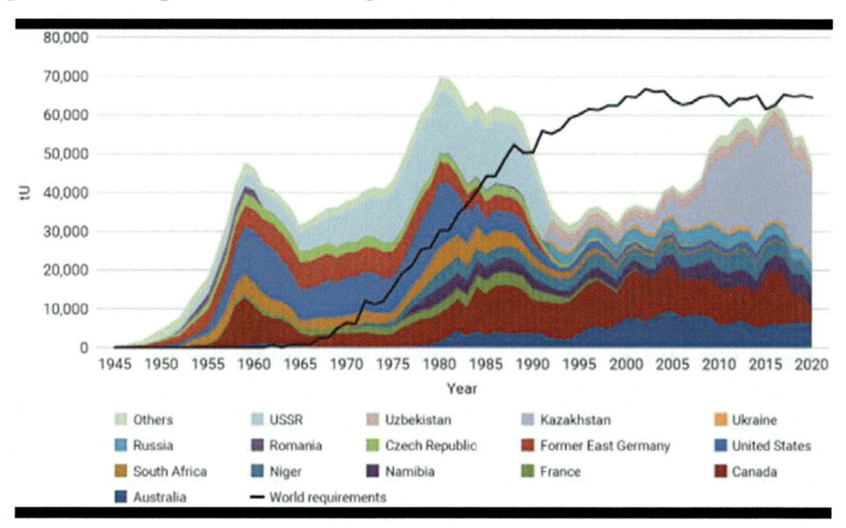

Several countries are extracting uranium ore and producing fuel for their nuclear power plants. The countries that are exporting the ore are important in the overall economy of the nuclear field. The various risks of reducing or even stopping exports are compensated by strong measures of the nuclear power plants countries to make large reserves of U ore or already made fuel. The future of the international price of U-ore depends on several factors, such as the development of new power plants, political risks, and market risks in exporting countries. The development of thorium reactors should change the nuclear power balance since thorium reserves are quite large, and the developed technology may cover thousands of years of consumption in countries that have reserves and implement the technology. Brief data on nuclear technologies, especially small modular reactors (SMRs) and thorium reactors, are provided in the technological resources.

It is clear that if we want to have a real circular economy, a common conscience about a planetary-level approach to managing resources with global cooperation must be established for using technological knowledge with the final goal of reducing environmental temperature increases as just one component in the complex process of reaching a closed thermodynamic system relative to the environment. This goal must include both resource management and technology development; some suggestions are given in later discussions herein.

3.7 Technological resources

The need for raw materials is related to the level of technological development of the economy. New technologies may serve to recycle waste, thus turning liabilities into assets. Moreover, in the energy field, various technologies have penetrated the niche of technologies and changed the way we interact with nature. Some examples are detailed below from the USEA handbook on such technologies (USEA/USAID Handbook of Climate Change Mitigation Options for Developing Country Utilities and Regulatory Agencies, June 1999).

3.7.1 Waste heat recovery system characteristics

In the process of electricity generation, only a portion of the available energy can be converted into useful energy. A significant portion of the heat energy available in the combustion of fuel is wasted, as demonstrated by operating efficiencies in the range of 34%; today's most efficient combined-cycle systems have average efficiencies in the low 50% range. The largest source of waste heat is the warm water produced by steam condensation. This waste heat can be used for process steam, hot water heating, space heating, and other thermal needs. If the energy content is sufficiently high, the steam could also be used to generate additional electricity in a cogeneration (or combined heat and power) system.

Current water heating system design methods focus on meeting hot water needs and generally ignore energy consumption, operating costs, and other effects. As a result, utility customers often fail to make informed decisions and consequently sacrifice potential savings and benefits. It is possible to funnel this steam through a steam turbine to cogenerate electricity. Doing so can improve the efficiency of the system to as much as 80 or 85%. Using the heat from this source to generate electricity could displace fossil fuel consumption, thereby avoiding greenhouse gas (GHG) emissions.

Hot flue gases from boilers can provide a source of waste heat for a variety of uses. The most common use is for preheating boiler feed water. Heat exchangers used in flues must be constructed to withstand the highly corrosive nature of cooled flue gases.

SIZE: Wide range of sizes from <25 to 300 MW.

FEATURES: Water temperature above 60–82°C (140–180°F) is required for domestic applications. Some equipment also acts as a silencer to replace or supplement noise reduction equipment needed to meet noise requirements.

COST: Installation costs can be $1000/kW (for industrial systems).

CURRENT USAGE: Waste heat recovery is used extensively in Central Europe and at industrial facilities around the world.

The City of Renville, Minnesota has implemented a waste heat recovery system in which an effluent hot waste stream (6000 gallons of water/minute @ 90–120°F) is pumped to one of two hot water heat exchangers. The heat from the hot waste stream is captured by a heat exchanger; the cooled waste stream is returned to a plant for further cooling and treatment. The captured heat is transferred to water in a closed loop that travels 4000 feet to the industrial user, in this case, a fishery (other users may include greenhouses or hydroponic farming, for example). After the heat is extracted for use, the water in the closed loop is reheated and recirculated. The present system of four pumps and two heat exchangers can be doubled. As presently configured, the project provides 35 million Btus from 2000 gallons per minute of hot water circulating through a 24-inch diameter pipe system. The total cost of reconfiguration was approximately $657,000 part of which will be repaid by industrial park user fees. http://www.bolton-menk.com

POTENTIAL USAGE: With increasing industrial markets worldwide, this represents an opportunity for a relatively low-cost increase in power capacity.

3.7.1.1 Issues with implementing action

Waste heat streams from the electricity generation process are not readily adaptable for conventional heating or process use.

A lack of space may prohibit the establishment of a waste heat application.

Need for a backup heat supply during outages.

It may be difficult to find a thermal host able/willing to locate near a power plant.

3.7.1.2 Climate change impact

EMISSION EFFECT: AVOIDED

3.7.1.3 Conditions for emissions mitigation

Emissions will be reduced directly because of increased efficiency of generation and distribution; and indirectly through reduced electricity demand.

In many instances to date, waste heat is used to replace electric capacity from retired, less-efficient units.

EMISSION ESTIMATE: To the extent that the overall system improves efficiency, GHG emissions will be reduced.

COST-EFFECTIVENESS: N/A

SECONDARY EFFECTS: NO_x and SO_2 reductions will depend on the generation mix.

3.7.1.4 Resources

The Electric Power Research Institute has sponsored the development of HOTCALC, a microcomputer software program that simulates the performance of commonly available commercial water heating systems to provide information for applications and design.

Thermal Energy Storage for Process Heat and Building Applications, SERI/TR-231 1780, http://www.epri.com.

Center for the Analysis and Dissemination of Demonstrated Energy Technologies, Heat Exchangers in Aggressive Environments, Analysis Series # 16, 1995.

Goldstick, R. and A. Thumann, 1986, Principles of Waste Heat Recovery, The Fairmont Press, Inc. Provides information about recovering heat at low, medium, and high temperatures for reducing operational costs.

The US Department of Energy provides a reference brief on heat recovery in commercial buildings, which includes a reference list of additional information. http://www.eren.doe.govconsumerinfo/refbriefs/ea4.html

3.7.2 End-use energy efficiency and demand side management actions

Demand-side management projects are designed to reduce energy consumption at the consumer level while maintaining the same level of energy services as before project implementation. There are two types of DSM projects:

Energy Efficiency (EE): focuses on modifications in end-use technology (e.g., lighting).

Energy Conservation: focuses on changing energy consumption use patterns through educational projects and the use of time-of-day tariffs.

Many DSM projects involve a combination of both energy efficiency and energy conservation measures that can result in low- and no-cost climate change mitigation options. The level (and cost) of reduction is dependent on the source of electricity. If the electricity was generated by fossil fuels (e.g., coal, oil, natural gas), then the reduced demand would translate into less generation and emission of GHGs.

However, the supply and corresponding emissions response is not always straightforward. Reduced electricity demand could cause a price adjustment and a "rebound effect"—where the reduced demand is offset by increased demand elsewhere on the system so there is no net change in generation or GHG emissions, or there is a realignment in the sources of generation so that even with reduced electricity demand, lower GHG emissions may not be realized. Quantifying the environmental impacts of EE/DSM actions should involve a detailed, system-specific analysis that recognizes the magnitude as well as the timing of the actions on the resultant electricity savings and generation capacity of the utility.

This section discusses some of the EE/DSM actions that are available to utilities and regulators to reduce GHG emissions. Estimates of the GHG emissions reduced or avoided by these actions are provided where available. However, since these actions are dependent on the mix of fuels/technologies used to generate the electricity being displaced, a more precise estimate of carbon emissions reduced/avoided requires site-specific details.

Some of the information sources used to prepare the discussion of EE/DSM actions, and in which more detailed discussions exist, are:

International Directory of Energy Efficiency Institutions, World Energy Efficiency Association (WEEA). A searchable version is also available online at http://www.weea.org. The WEEA site also hosts the Energy Efficiency Technical Library and case studies of energy efficiency projects by country and region.

The Center for the Analysis and Dissemination of Demonstrated Energy Technologies (CADDET), based in the Netherlands, provides information (and case studies) on energy efficiency and renewable energy technologies installed in industrialized nations. The CADDET site, http://www.caddet-ee.org, includes copies of newsletters, announcements, and events, and a publications catalog.

The Impact of Global Power Sector Restructuring on Energy Efficiency, and accompanying Bibliography, Hagler-Bailly, prepared for USAID Office of Energy Environment, and Technology Reports No. 98-02 and 98-05 (1998). The Bibliography provides a detailed listing of authoritative resource documents and articles on energy efficiency, including the "Top 25 Key Reference Documents," and publication and distribution information.

The International Institute for Energy Conservation posts several project summaries and other guidance documents at http://www.iiec.org.

The Alliance to Save Energy has a variety of programs promoting efficiency in buildings and industry, energy policy reform, education and outreach, and NGO development. Information on the Alliance's initiatives is found at http://www.ase.org.

The American Council for an Energy-Efficient Economy (ACEEE) links to industry conferences, consumer information, and other websites, as well as information on ACEEE publications. http://www.aceee.org/index.html.

The National Association of Regulatory Utility Commissioners published Incentives for Demand-Side Management in October 1993. This report profiles US state commission policies and activities encouraging utility investments in demand-side management resources, including a table that provides a quick summary of each state commission's DSM activities, with each type of shareholder incentive mechanism summarized. These and other documents can be found at the NARUC website http://www.naruc.org.

The International Energy Agency hosts a website dedicated to the compilation of summaries of demand-side management and energy efficiency efforts around the world. The site provides country-by-country information on electric power restructuring, existing mechanisms, and any issues affecting the uptake or success of EE/DSM programs. http://dsm.iea.org.

The US EPA (Region 4) provides an energy efficiency action plan checklist for companies interested in improving their energy efficiency at: http://www.epa.gov/region04/air/caileea.htm.

Lawrence Berkeley National Laboratory's (LBNL) Energy and Environment Division has developed several energy-efficient technologies and technical support services. Information on these products and services is provided on the LBNL website at: http://eetd.lbl.gov/.

3.7.3 Promoting residential demand-side management programs

3.7.3.1 Characteristics

The IPCC estimates that by 2030, residential buildings will account for approximately 60% of energy use in buildings, regardless of the country.

Therefore, reducing energy use in residential buildings will have a significant impact on future GHG emissions. By reducing (or avoiding) energy and electricity use, residential DSM programs will avoid the associated GHG emissions that would otherwise have been produced.

Residential DSM activities encompass a broad range of utility/customer interactions. Activities typically involve energy conservation, or "load shaping"/"load shifting" methods to reduce peak demand requirements by encouraging (l) the installation of energy-efficient equipment or (2) voluntary reduction in consumption through behavioral adjustments.

SIZE: Scalable to any size residential building.

FEATURES: Residential DSM programs are primarily directed at (1) improving the energy efficiency of customer appliances (heating. cooling and lighting), (2) improving the energy efficiency of new and existing construction (through weatherization and design), and (3) managing residential load (home automation systems, rate structures that encourage off-peak electricity consumption, and other tools).

COST: Costs vary by program and can be minimal (load shifting) or involve the installation of new computer equipment or other hardware requiring significant upfront costs (although these costs can be recovered/ saved over the lifetime of the equipment).

CURRENT USAGE: Many OECD countries have achieved considerable energy savings through DSM programs. Non-OECD countries with significant DSM programs underway include Brazil, Pakistan, Thailand, Mexico, Jamaica, and the Philippines.

POTENTIAL USAGE: One estimate is that DSM can reduce worldwide energy demand by 5%–6% by the year 2050. Conservation programs are applicable in every country. Implementation is limited only by the willingness of utilities to engage in the programs.

3.7.3.2 Issues with implementing action

A lack of uniformity in residential building energy codes and retail energy efficiency standards contributes to the difficulty of producing energy-efficient residences and technologies.

Inadequate information on the costs and benefits of residential DSM activities contributes to the lack of familiarity and information on the installation of energy-efficient technologies.

There is a lack of capital for customer purchases of new, high-efficiency equipment. Incentives from utilities, including financing, can reduce the cost of energy-efficiency equipment and encourage participation in load management and conservation programs. Utilities can increase the direct installation of cost-effective conservation measures by targeting applications where customers lack sufficient motivation and/or resources. There is no real-time pricing for residential customers.

3.7.3.3 Climate change impact

EMISSION EFFECT: REDUCED

3.7.3.4 Conditions for emissions mitigation

The level of GHG emissions decreased or avoided will depend on the technologies used and the generation mix before and after the energy efficiency program For example, improving the efficiency of refrigerators—or other appliances that operate continuously—would result in larger energy savings than efficiency improvements in peak technologies, such as air conditioning units. However, if baseload power is provided by a clean source (e.g., hydro) and peak power by a fossil-fuel-powered source, then improving the efficiency of the peak appliance would have a greater effect on emissions. conditioning, water heating, and miscellaneous electricity usage. The greatest energy and carbon savings can be achieved in lighting, and space magnitude of its DSM programs.

EMISSION ESTIMATE: Varies according to the change in electricity demand before/after the implementation of the DSM program.

COST-EFFECTIVENESS: Varies according to the administrative or investment costs required. Some investments are cost-effective regardless of the energy savings achieved.

SECONDARY EFFECTS: Varies according to the decrease in electricity demanded. For every kWh of fossil fuel power generation avoided, the associated emissions of air pollutants are also avoided.

3.7.3.5 Examples

The Ilumex project in Mexico helped consumers purchase compact fluorescent lights using an innovative financing scheme that allowed consumers to purchase the lamps with a loan that could be repaid from electricity bill savings. More than 600,000 lamps have been sold to date for US$1.64 each, resulting in estimated annual energy savings of 160 GWh/year, 34,400 tons C, and 2510 tons SO_2. The program has also allowed the Mexican utility to avoid the construction of 78 MW of new peak generating capacity.

To quantify the level of emissions reductions, a utility can use a planning and dispatch model (or production cost model) to identify planned electricity dispatch and estimate the load shape.

3.7.3.6 Resources

The US Department of Energy has sponsored a variety of residential DSM programs, including:

1. The Cool Communities Program, which develops community partnerships to plant trees and increase the use of highly effective exterior surfaces on buildings and roads to reduce heating and cooling costs and improve the environment
2. The National Earth Comfort Program, an industry/government partnership to increase the geothermal heat pump market from 40,000 to 400,000 units/year
3. E-Seal, a national initiative to promote energy efficiency and environmental awareness in both new construction and retrofit home programs
4. The Office of Building Technology website, containing links to technical information, case studies, and other background information promoting residential energy efficiency, at http://iwww.eren.doe.gov/buildings

The US Environmental Protection Agency has also sponsored several residential DSM initiatives. These include:

1. Energy Star Programs to encourage the production and use of energy-efficient equipment. http://www.epa.gov/energy star
2. Green Lights, whose more than 1650 participants have invested in efficient lighting; those involved have reduced their lighting electricity consumption by an average of 47%, saving approximately $90 million.

Power Smart, a brochure published by the Alliance to Save Energy, identifies a range of residential DSM actions that homeowners can take to conserve energy and use energy-efficiently. http://www.ase.org. ACEEE produces the Consumer Guide to Home Energy Savings, a directory of the most efficient products available.

3.7.4 Promoting commercial demand-side management programs

3.7.4.1 Characteristics

Commercial DSM projects are sponsored by utilities to encourage commercial customers to control their energy bills, and simultaneously enable the local utility to achieve its own load goals. In the United States, commercial buildings-including commercial construction and multi-family residential buildings consume 11% of total energy. Energy is used in commercial buildings to provide a variety of services such as lighting. space heating and cooling, refrigeration, and electricity for electronics and other equipment. Commercial DSM projects encourage builders and occupants of commercial spaces to increase both energy conservation and energy efficiency.

Analysis indicates that substantial reductions in future GHG emissions can be realized through the use of more energy-efficient technologies. In addition to avoiding GHG emissions, energy-efficient technologies also can improve indoor air quality, reduce noise, improve process control, and increase amenities or convenience.

Current commercial DSM initiatives target heating and cooling systems, lighting, and engineering systems. Commercial DSM programs can include (1) technical and financial assistance for local efforts to promote commercial efficiency partnerships between manufacturers, utilities, and end-users to develop highly efficient equipment, (2) grants and training to local officials to update commercial building codes, etc. Such programs lower operating expenses, improving cash flow for building owners. In commercial buildings where occupants pay utilities, energy-efficient buildings are more attractive to potential tenants. Energy costs can be reduced by as much as 50% with the installation of efficient lighting, space conditioning, and building controls.

SIZE: Can be adapted to any size building.

FEATURES: Can include building envelopes; efficient equipment (e.g., lighting, motors, variable speed drives, HVAC equipment): thermal storage equipment; subsystems control; load management; and community energy systems (district heating and cooling).

COST: Varies with the program. It can be minimal (shifting usage) or can involve installation of new computer equipment, etc. that may require upfront capital investment, but can be recovered through energy savings over the system's lifetime.

CURRENT USAGE: Organizations in OECD countries are very active in promoting commercial DSM programs. Non-OECD countries with significant DSM programs underway include Brazil, Pakistan, Thailand, Mexico, Jamaica, and the Philippines.

POTENTIAL USAGE: One estimate is that DSM can reduce worldwide demand by 5.6% by the year 2050. To achieve these targets, DSM programs must be universally applied.

3.7.4.2 Issues with implementing action

There is a lack of customer and designer awareness of energy-efficient technologies many products that can reduce GHG emissions are relatively new or are not the standard approach to design. Designers tend to specify what they know and have used in the past. Providing product information, design assistance, and public seminars/training programs may increase familiarity with technology.

It may not be the decision-maker who pays the utility bill. A high percentage of commercial space (office, retail, etc.), is leased rather than owned by the occupant. Price signals, DSM incentives, and economic paybacks are distorted by disconnects, between building ownership, energy usage, and responsibility for energy bills.

There is a lack of available energy-efficient equipment in the marketplace, and what is available usually has a higher initial cost relative to conventional technologies. The technical potential for improved electric products exists but will not be produced until manufacturers are assured of sufficient demand.

Conflicting priorities influence investment decisions, and investment in energy efficiency improvements usually compete with esthetic improvements, employee productivity investments, comfort or safety concerns, product integrity protection am other Concerns for commercial business owners. In some cases, energy efficiency improvements not only compete for capital expenditures but also may be viewed as a trade-off that requires a sacrifice in comfort or esthetics.

3.7.4.3 Climate change impact

EMISSION EFFECT: REDUCED

3.7.4.4 Conditions for emissions mitigation

The greatest carbon (and energy) savings are achieved in space conditioning, lighting, and miscellaneous electricity use.

EMISSION ESTIMATE: Varies according to the change in electricity demand before/after implementing the DSM program.

COST-EFFECTIVENESS: Varies according to the administrative or investment costs required. Some investments are cost-effective regardless of the energy savings achieved.

SECONDARY EFFECTS: Varies according to the decrease in electricity demand. For every kWh of fossil fuel power generation avoided, the associated emissions of air pollutants are also avoided.

To quantify the level of emissions reductions, a utility can use a planning and dispatch production cost model to identify planned electricity dispatch and estimate the load shape and magnitude of its DSM programs.

3.7.4.5 Resources

Green Buildings for Africa program, South Africa. The Council for Scientific and Industrial Research (CSIR) launched this voluntary program in June 1997 to encourage building owners to improve energy efficiency. Participants commit to assess their needs/options within 6 months and make upgrades within 3 years. CSIR provides technical assistance, and participants earn the right to use the program logo in their advertising. This program is expected to reduce building energy consumption by 30% over 3 years.

United Nations Environment Programme, 1997, *Reducing Greenhouse Gas Emissions: The Role of Voluntary Programmes.*

Koomey, J.G. et al., 1994, Buildings Sector Demand-Side Efficiency Technology Summaries, Lawrence Berkeley National Laboratory, LBNL-33887.

Lawrence Berkeley Laboratory has worked with the US Agency for International Development to support efforts to implement existing energy standards in new commercial buildings in the Philippines.

The US Environmental Protection Agency sponsors the Energy Star Buildings and Green Lights Programs whose participants agree to install energy-efficient lighting where profitable as long as lighting quality is maintained or improved. http://www.epa.govgreenlights.html/

The US Department of Energy Office of Building Technology website http://www.eren.doe.gov/buildings contains links to technical information, case studies, and other background information on energy-efficient technologies appropriate for commercial use.

3.7.5 Promoting industrial demand-side management programs

3.7.5.1 Characteristics

Industrial DSM programs target one of the four groups of primary end-use electricity applications: motor drives and controls; process applications; lighting; and space conditioning.

Currently, industrial energy use produces almost half of global CO emissions, yet energy intensity varies dramatically. In many developing countries, energy intensity is two-to four times greater than the average in OECD countries. Improvement in industrial energy efficiency can potentially have a tremendous impact on reducing GHG emissions.

Industrial DSM activities can encompass a broad range of utility/customer interactions including load management, interruptible rates, time-of-use pricing, and end-use applications. Electricity reductions can be achieved through fuel switching, cogeneration, and process energy efficiency improvements. Results of industrial DSM programs can include reduced capital requirements; reduced energy consumption; reduced spoilage/waste; flexibility of raw material base: greater control over production; improved product quality and yield; decoupling of production from fuel supply; increased competitiveness; increased production at less energy per unit produced; and reduced environmental impacts (net and/or site).

SIZE: Programs can target individual facilities, specific industries, or the entire industrial sector.

FEATURES: In the United States, the industrial sector accounts for approximately 40% of energy and electricity consumption. This large industrial market represents major opportunities for large load increments of load growth, conservation, and management.

COST: Varies by program. It can be minimal (shifting usage) or can involve the installation of new computer equipment to manage load, etc.

CURRENT USAGE: In addition to OECD countries, countries such as India, Senegal, Georgia, Jamaica, and Mexico have successfully employed DSM programs to reduce industrial demand.

POTENTIAL USAGE: One estimate is that DSM can reduce worldwide electricity demand (from business-as-usual) by 5%–6% by the year 2050. Significant emissions reductions to date have been experienced in OECD countries in the chemicals, steel, aluminum, cement, paper, and petroleum refining industries. *The industrial sector includes all manufacturing, agriculture, mining, and construction activities.

3.7.5.2 Issues with implementing action

Existing capital has a long life; equipment upgrades may not be needed for several years.

At many plants, there may not be anyone directly responsible for energy efficiency; and decision-makers are often located far from regional plant sites.

Modifications to industrial manufacturing processes are complex and can affect and limit coordination with facility shutdowns, energy cost savings, product quality, and productivity. The high cost of process modifications alone may prevent them from meeting industry payback and return-on-investment requirements.

Energy efficiency expenditures are not competitive on a rate-of-return basis with product improvement expenditures. Additionally, electricity is a small part of many companies' overall costs so they may dismiss potential energy cost savings as insignificant.

It is difficult to standardize industrial DSM applications because many companies consider their production processes proprietary, and need assurances that their operations will be kept confidential.

3.7.5.3 Climate change impact

EMISSION EFFECT: REDUCED

3.7.5.4 Conditions for emissions mitigation

The amount of emissions decreased or avoided will depend on the technologies needed before and after the energy efficiency program and the generation mix.

EMISSIONS ESTIMATE: Varies according to the change in electricity demand before/after implementation of the DSM program.

COST-EFFECTIVENESS: Varies according to the administrative or investment costs required. Some investments are cost-effective regardless of the energy savings achieved.

SECONDARY EFFECTS: Varies according to the decrease in electricity demand. For every kWh of fossil fuel power generation avoided, the associated emissions of air pollutants are also avoided.

3.7.5.5 Resources

The Industrial Energy Efficiency Network (IEEN) was established by the Norwegian government in 1989 to promote energy efficiency. IEEN works with individual companies in 13 industrial sectors to disseminate information, provide technical data on new technologies, and compile statistics on energy usage. Results have varied within industrial sectors, but many companies have achieved from 10% to 50% reductions in energy consumption. http://www.ife.no/departments/energy/lindex.html

To quantify the level of emission reductions, a utility can use a planning and dispatch production cost model to identify planned electricity dispatch and estimate the load shape and magnitude of its DSM programs.

The US Department of Energy has sponsored the following industrial DSM programs:

1. Motor Challenge to promote voluntary collaborative efforts between the private and public sectors to demonstrate, evaluate, and accelerate the use of efficient industrial electric motor systems.

2. National Industrial Competitiveness for Energy, Environment and Economy, which awards grants to improve industrial process efficiency, reduce waste, and cut GHG emissions in several key industries.

Brazil's PROCEL, the national electricity conservation program, developed demand-side management projects saving 250 MW in electricity and US$500 million in power plant development costs.

Mexico's federal electricity conservation program, Fideicomiso de Apoyo al Programa de Ahorro de Energia del Sector Electrico, conducted innovative energy efficiency demonstration projects that brought a 5% decline in projected energy use.

Studies have concluded that in China alone, raising the efficiency of industrial furnaces—which consume about 259% of China's energy production—would reduce the energy used by furnaces by 40% and avoid the waste of about 2.7 quadrillion Btus (2.9 EJ) per year.

Lawrence Berkeley Laboratory has worked with the US Agency for International Development to support efforts to develop motor standards in the Philippines. Information on the European Conference on Industrial Energy Efficiency can be found at http://www.eva.wsr.ac.at/indeff/prog-e.htm.

The Alliance to Save Energy provides links to more information on industrial energy efficiency at http://www.ase.orgprograms/industrial/links.htm.

The European Union (DG-XVII) sponsors several energy-efficiency programs, providing background information and funding. http://www.europa.eu.int/en/comm/dgl7/programs.htm

3.7.6 Renewable energy actions

Renewable energy is most often defined as energy derived from inexhaustible sources—the sun, the wind, and the earth. In addition, renewable energy options often include the use of wastes to produce high-valued and useful energy. The use of renewable energy has the advantage of producing little or no carbon emissions. Wind power, photovoltaics (PVs), biomass derived from continuously replenished sources, and hydroelectricity have no net carbon emissions. Although energy derived from wastes can emit carbon, such emissions are usually no more than for natural gas-based systems. As a result, using renewable energy in lieu of fossil-based systems would significantly lower carbon (and GHG) emissions. Approximately 30 quadrillion Btus of renewable energy are currently consumed throughout the world. This represents approximately 8% of world energy consumption. Although the large majority of renewable energy consumption is from large hydroelectric sources (and in some countries, biomass), the use of biomass, wind, and PVs is growing rapidly. For example, it is estimated that, for each of the last few years, more than 1200 MW of new wind generating capacity have been installed throughout the world and the figure is increasing. Significant market opportunities for renewable energy exist. This section provides information on the factors that will influence the viability of each type of renewable energy. It also provides some information on sources for financing renewable energy projects.

3.7.7 Biomass

3.7.7.1 Characteristics

Biomass energy utilizes the energy content of energy crops, agricultural residues, wood wastes, or animal wastes. or gasified. Direct biomass combustion technology is quite similar to coal combustion technologies, with materials that are either combusted in boilers to produce steam or heat or used fairly easily to "cofire" biomass with coal in existing boilers. Biomass can also be converted into combustible gases, much as natural gas is used to generate electricity or fuel vehicles.

When used to offset fossil fuel use, bioenergy systems can significantly reduce or eliminate GHG emissions. While some GHG emissions may be produced through the combustion of biomass for electricity, total emissions/kWh are considerably less than those produced by fossil fuels. Additionally, if a replacement crop is planted, equivalent to the amount of biomass used for electricity generation, the hew crop absorbs the CO, produced by the combustion process resulting in no net emissions. In the world, over two billion people are without electric power, and another billion have less than 5 hours a day of electricity. Most of these three billion people live in the middle latitudes where biomass grows abundantly and can potentially be a viable source of fuel. Advances in biomass combustion technology contribute to its increasing viability. Further development of biomass gasification technology is ongoing, with expectations that the next generation of technologies will reach efficiencies of 65%–70%. Conversion technologies for power plants that co-fire biomass with coal or municipal waste are also being developed.

SIZE: 2–100 MW, average size is ~20 MW

FEATURES: Peaking power and baseload applications (>6000 h/year) with 15–30% efficiency; cogeneration applications can reach 60% efficiency.

COST: Costs will vary according to local conditions, but as a guideline: \$530–600 kW for industrial units and \$300 kW where fuel sources are geographically convenient.

CURRENT USAGE: In the United States, installed biomass capacity for electricity generation is over 6.5 GW (over 3% of US energy consumption). In Finland, Sweden, and Austria, 13%–18% of electricity generated is fueled by biomass.

POTENTIAL USAGE: Resource and market assessments identify an extremely broad range of potential, with the greatest potential in developing countries. By 2050, estimates indicate that biomass could provide 17% of the world's electricity and 38% of direct fuel use.

These include wood, bark, agricultural residues, peat, and refuse-derived fuels (with 50% moisture). All of these can be used in their raw form or can be pelletized through drying and compressing biomass materials. When pelletized fuels are used, average efficiencies reach 70%–80%.

3.7.7.2 Issues with implementing action

A large, steady supply of biomass is required for reliable electricity generation.

Biomass supply may be climate or season dependent.

Land suitable for biomass development may face competition for other uses and/or there may be opposition to harvesting existing resources such as forests

The cost of procuring feedstock may be prohibitive where biomass must be transported long distances to a combustion site. Since biofuels have a relatively low energy content per ton, bioenergy facilities must be sited close to their fuel Source to minimize transportation costs. However, co-firing biomass/coal may stabilize the fuel supply for such plants.

Typically, biomass contains 1%–4% noncombustible ash by weight, which may require special disposal arrangements. Such ash often contains low levels of lead, barium, selenium, and arsenic and must be carefully landfilled.

3.7.7.3 Climate change impact

EMISSION EFFECT: OFFSET

3.7.7.4 Conditions for emissions mitigation

Biomass used to produce energy can avoid a net increase of CO in the atmosphere if it is replaced by new growth that absorbs an equivalent amount of CO. Total emissions will vary according to the boiler/combustor system used.

AVOIDED EMISSION ESTIMATE: 200 MtC/MW_e/year offset

COST-EFFECTIVENESS: Estimated net cost of CO, avoided is from $25–38/ton.

SECONDARY EFFECTS: May produce some methane (CH). As with carbon emissions, when biomass is used to offset fossil fuel use, bioenergy systems can significantly reduce or eliminate S0, NO, and particulate matter.

3.7.8 Geothermal

3.7.8.1 Characteeristics

Geothermal energy channels the heat and steam stored below the earth's surface to a turbine that then drives a generator to produce electricity. Underground heat sources such as hot water reservoirs can be tapped by drilling through the earth's layers. Some surface manifestations such as hot springs and geysers may also be tapped for electric generation. High-temperature steam sources are of the greatest value for generating electricity; geothermal plants operate over temperature ranges of 122–482°F (50–250°C), a relatively low heat compared with traditional fossil or nuclear plants. As long as resources are sustainably managed, geothermal can serve a baseload power-generating function with high availability. The rate at which the hot water or steam is replenished—either naturally or by injecting spent fluids—determines the quantity of energy that the source is capable of supplying continuously.

Geothermal energy produces minimal amounts of carbon dioxide and only traces of nitrogen oxide and sulfur dioxide emissions. Closed-loop Systems, the newest generation of geothermal technologies, produce no airborne emissions. As a generator of baseload electricity, geothermal competes with fossil fuel power Sources. electricity generated by a geothermal site avoids the emissions that would have otherwise been produced by combusting fossil fuels

For power generation, geothermal is limited to site-specific availability, and locations may not be close to a transmission grid. However, there is tremendous untapped potential for developing geothermal resources in Asia, and new technologies under development will improve the economics of using smaller geothermal sites for power generation, increasing the number of sites with economic potential. Of the five forms of geothermal energy, only two—hydrothermal reservoirs and earth energy—are currently used for electric power generation. Technological advances must be made before the three other forms—geopressured brines, hot dry rock, and magma—can be commercially developed.

Thus, every kW at temperatures under 90°C is used for heating buildings and process heat.

SIZE: 1–110 MW

FEATURES: Geothermal power plants are highly reliable and can operate 24 h/day. Average availability is 80%, but many plants have >95% availability.

COST: $840–$2500/kW

CURRENT USAGE: 6300 MW installed in 21 countries worldwide. Direct use of geothermal water occurs in >40 countries.

POTENTIAL USAGE: An additional 6000 MW is potentially economically viable worldwide. Worldwide resource potential >40,000 MW.

3.7.8.2 Issues with implementing action

Availability of appropriate sites is limited and distributed unevenly.

Production must be carefully managed if the resource is to remain sustainable

Initial capital and development costs are high: special materials and construction techniques are required to mitigate the erosion and corrosion caused by water and steam

Gases such as H and S may be emitted from geothermal wells, but developers can collect and reinject the gases.

Cost per kWh is competitive with coal and nuclear power, but geothermal is not yet competitive with new NGCC technologies (where natural gas is readily available)

3.7.8.3 Climate change impact

EMISSION EFFECT: AVOIDED

CONDITIONS FOR EMISSIONS MITIGATION: Geothermal power plants meet the most stringent environmental regulations and release small amounts of CO_2. With advanced, closed-loop technologies (binary geothermal technology applied to resources with a temperature below 350°F or 177°C), no emissions are produced.

EMISSION ESTIMATE: 56 MtC/GWh avoided. The newest generation of flash-steam plants emit only 1lb/CO_2/MWh_e.

COST-EFFECTIVENESS: Net cost is $0–144/ton of CO_2 avoided.

SECONDARY EFFECTS: Sulfur emission rates range from zero to a small percentage of those produced by fossil fuels. traces of nitrogen oxides. Produces some H_2S.

3.7.9 Small-scale hydropower

3.7.9.1 Characteristics

Small-scale hydroelectric systems (under 20 MW) capture the energy in flowing water and convert it to electricity. Systems can be either "run-of-the-river" or pumped storage* and are suitable for stand-alone (isolated) or grid-connected applications.

If they are well-designed, small hydroelectric systems blend with their surroundings and have minimal negative environmental impacts. As with larger hydroelectric systems, small hydro produces no GHG emissions.

The potential for small hydroelectric systems depends on the availability of suitable water flow. Where the resource exists, the development can provide cheap, clean, reliable electricity. Locations are numerous around the world and are often accessible to load centers and the grid.

SIZE: 1–20 MW

FEATURES: Operating Efficiency: 85%–88%. Capacity factors vary from 20% to 90% depending on the variability in streamflow. To produce 200 W, water sources must have, at a minimum, a change in elevation (or head) of 20 feet @ 100 gallons/minute (or 100 feet of head @ 20 gallons/minute). Areas with a low head will need long runs of large-diameter pipe. Additionally, distances of over a few hundred feet may require the construction of expensive cabling

COST: $1.000–3000/kW. Costs vary widely based on site-specific factors such as streamflow, geological characteristics, and the extent of existing civil structures at the site. Major costs ara associated with site preparation and equipment purchase.

CURRENT USAGE: As of 1993, 20% of global electricity was generated by hydro: small-scale hydroplants of 10 MW or less account for 4% of total hydrogeneration.

POTENTIAL USAGE: The continents of Africa, Asia, and South America have the potential for 1.4 million MW, four times as much capacity as is currently built in North America. Less than 10% of the world's (total, large, and small) technical useable hydropower potential is being used today.

The plants may be even smaller, or microhydro (2–300 W) systems are appropriate for residential applications.

Run-of-the-river hydroelectric plants use the power in river water as it passes through the plant without causing an appreciable change in the river flow. Normally such systems are built on small dams that impound little water or may be built without any reservoir or dam. Pumped storage projects provide a means of storing energy. Excess off-peak energy is used to pump water to an upper reservoir where it is stored as potential energy. The water is then released to produce peak-load power when necessary.

3.7.9.2 Issues with implementing action

Availability of resources is site specific and may not be located close to demand centers.

3.7.9.3 Climate change impact

3.7.9.3.1 Emission effect

Conditions for emissions mitigation Hydropower produces no GHG emissions, Environmental impact may occur due to land-use or siting issues,

EMISSION ESTIMATE: Produces no GHG emissions.

COST-EFFECTIVENESS: \$25−38/ton of net CO_2 avoided

SECONDARY EFFECTS: Produces no air pollutants.

3.7.10 Photovoltaics

3.7.10.1 Characteristics

Solar cells are thin wafers of silicon or other semiconductor materials that create an electric current when sunlight strikes them. Solar cells are the building blocks of a PV system. Current commercial PV devices are mounted on a surface and wired in series to become solar modules, with a maximum conversion efficiency of about 15%. Direct sunlight, as well as diffuse light scattered by clouds or humidity, can generate electricity, making PV an option in warm and cool climates. PV is by nature an intermittent resource, but where it matches daytime power peaks, it may be economical for utility applications. Additionally, the solar PV module's relatively high initial cost (currently ~\$5/watt) is offset by a very long life (as much as 30 years) and relatively low maintenance requirements.

PVs produce no net emissions, although some toxic chemicals are used to produce PV systems. Where PVs are used for baseload or peaking power, fossil-fuel-generated electricity is avoided.

PV modules can be used without moving parts; their support structures can be fixed in place or designed to track the sun across the sky. Large-scale PV power plants, consisting of many PV arrays installed together, can prove useful to utilities for many reasons. Utilities can build PV plants much more quickly than they can build conventional power plants because the arrays are easy to install and connect. In addition, because siting PV arrays is much easier than siting conventional power plants, utilities can locate PV plants where they are most needed by the grid. Finally, unlike conventional power plants, PV plants can be expanded incrementally as consumer demand increases. New developments include increased production capacity and improved technological efficiency to reduce costs.

SIZE: Modules range from a few watts to multiple megawatts. For power generation, modules can be combined to produce 5–10 MW or larger.

FEATURES: Maximum operating efficiency is 15% (sunlight-to-electricity), and average efficiency is 10%. Systems using trackers that follow the sun collect about 33% more sunlight than fixed arrays.

COST: Systems cost \$6000–20,000/kW, with a module cost of ~\$5000/kW, although expectations are that cost will decrease to \$1000/kWh and as low as \$700–800/kW by 2030. PV is competitive as a stand-alone power source in areas remote from electric utility grids. The levelized cost for large PV systems (>1 kW) is \$0.25–\$50/kWh, making PV cost-effective for residential customers more than a quarter mile (0.4 km) from the grid.

CURRENT USAGE: About 150 MW of PV is shipped every year, and more than 200,000 residential and commercial buildings use PV systems. demand is increasing at a rate of 15%–20% each year.

POTENTIAL USAGE: Solar insolation sufficient for PV exists in virtually every country in the world.

3.7.10.2 Issues with implementing action

Solar insolation varies geographically. Once PV equipment is purchased and installed, negligible additional costs are incurred. Fuel costs are zero, so PV systems may be more economical over a project lifetime. PV is becoming the power supply of choice for remote and small-power, DC applications of 100 W or less.

The cost of PV-produced electricity varies with atmospheric conditions—PV cells may lose 0.5% of their production efficiency for each degree Celsius above their rated temperature.

PV cannot provide continuous power without energy storage systems. Because of its variable nature (due to the variance of sunlight), utility planners must treat a PV power plant differently than they would treat a conventional plant.

3.7.10.3 Climate change impact

EMISSION EFFECT: AVOIDED

3.7.10.3.1 Conditions for emissions mitigation

In some applications, backup power generators (e.g., diesel) may be necessary.

EMISSION ESTIMATE: No direct GHG emissions; where backup power is necessary, some emissions would be produced.

COST-EFFECTIVENESS: \$26–400/ton of CO_2 avoided (net), depending on alternate fuel sources

SECONDARY EFFECTS: Produces no air pollutants although some systems involve the use of toxic materials which can pose risks in manufacture use and disposal.

3.7.11 Solar thermal

3.7.11.1 Characteristics

Unlike PV systems, which use sunlight to produce electricity directly solar thermal systems generate electricity with heat from concentrated sunlight collectors use mirrors and lenses to concentrate and focus sunlight onto a receiver mounted at the system's focal point The receiver absorbs and converts sunlight into heat The heat is then transported to a steam generator or engine where it is converted into electricity.

There are three main types of solar thermal electric systems: parabolic troughs, parabolic dishes, and central receiver systems.

(1) trough systems, or parabolic trough collectors, use mirrored troughs to focus energy on a fluid-carrying receiver tube located at the trough's focal point. There are 354 MW of trough systems installed in southern California;

(2) dish systems use parabolic mirrors to concentrate and focus incoming solar energy onto a receiver mounted above the dish at the focal point. Each dish produces 5–50 kW of electricity that can be used independently or linked together to increase generating capacity;

(3) central receiver systems, or "Power Towers," use thousands of individual tracking mirrors (heliostats) to reflect solar energy onto a receiver located atop a tall tower. The world's largest central receiver plant is a 10 MW power plant near Barstow California.

Solar energy technologies offer a clean, renewable, and domestic energy source. In the United States, solar thermal power plants produce about 480 million kWh of energy each year displacing 325,000 tons of CO, (6.8 tons of CO_2/kWh) annually.

Solar energy technologies have made huge technological and cost improvements, but except for certain niche markets—such as remote power applications-are still more expensive than traditional energy sources. Researchers continue to develop technologies that will make solar energy technologies-particularly power-generating technologies cost-competitive with fossil fuels. Current research efforts are focused on developing lighter, more efficient system components and energy storage technologies.

SIZE: Parabolic troughs: 30–80 MW, units now in deployment; 160 MW, modules proposed. Central receiver systems: 80–200 MW, Parabolic dish-Stirling engines: 5–50 kW,

FEATURES: Parabolic troughs: require ~2 ha of land per MW, of generating capacity, based on a daily mean direct normal insulation value of six to seven kWh/m^2. Central receiver systems: require 3–5 ha land/MW. Parabolic dish-Stirling engines: require about 0.7 ha/MW.

COST: Parabolic troughs: $2890/kW, (1991) Central receiver systems: $3300/kW–$100 MW, or $2800/kW–$200 MW (1991) Parabolic dish-Stirling engines: $1700–$3000/kW.

CURRENT USAGE: Over 354 MW, parabolic troughs, primarily in California.

POTENTIAL USAGE: Developing countries, where half the population is currently without electricity and sunlight is usually abundant, represent the biggest and fastest-growing market for power-producing technologies. The largest potential US application is for power production.

3.7.11.2 Issues with implementing action

Fuel source sunlight fluctuates seasonally and is nonexistent at night. For dispatchability, solar thermal technologies must have storage capacity or a backup power source, both of which increase overall system costs. However, resource availability does match daytime demand peaks.

Implementation requires large quantities of land for baseload or peaking electricity generation and may have a negative impact on habitats.

Reliability problems have arisen from the high temperature requirements and corrosive effects on solar mirrors.

Costs per kWh produced by stand-alone solar thermal sources are currently higher than with conventional sources.

3.7.11.3 Climate change impact

EMISSION EFFECT: AVOIDED

3.7.11.4 Conditions for emissions mitigation

EMISSION ESTIMATE: 6.8 tons/kWh of CO, avoided

COST-EFFECTIVENESS: \$88–178/ton of CO, avoided (net)

SECONDARY EFFECTS: Produces no emissions unless a hybrid natural gas system is used (parabolic troughs only) in which case emissions remain minimal except for NO_x @ 31.8 g/MWh.

3.7.12 Waste-derived fuels

3.7.12.1 Characteristics

A variety of waste products can be converted into liquid and gaseous fuels for use in generating electricity. These include municipal solid waste—which can also be processed into refuse-derived fuel that yields higher Btus and reduces the ash energy source—or biogas created from organic waste or biomass through fermentation. In addition, methane gas can be recovered from landfills and used in place of natural gas. Methane is a flammable gas produced from landfill wastes through anaerobic digestion, gasification, or natural decay.

Using these waste-derived fuels not only consumes what would otherwise be waste but also produces fuels with lower emissions than those associated with fossil fuels. Where methane is recovered, it avoids the release of methane, a GHG.

SIZE: 1–5 MW (modular combustion): 30–100 MW, (field-erected).

FEATURES: MSW is 25% efficient when used for power generation. Biogas can be used in internal combustion engines for shaft power or electricity; it can also be used as cooking and heating fuel. Methane recovery can achieve 70–80%o efficiency.

COST: \$91,000/Mt of daily capacity for MSW; \$4750/kW, for biogas. Methane recovery can be relatively inexpensive to operate, but if pipelines or other infrastructure is required, project costs increase dramatically.

CURRENT USAGE: Many European countries currently combust their waste streams; additional efficiency may be achieved through recovering heat or electricity. Currently, more than 100 power plants in 31 of the United States burn landfill-generated methane.

POTENTIAL USAGE: Applicable everywhere that waste is generated. Especially applicable in highly populated regions that produce significant quantities of waste.

3.7.12.2 Issues with implementing action

Creates ash as a by-product that must be disposed of, requiring a landfill. These disposal costs add to O&M costs.

Operators must have access to a significant quantity of garbage (100 TPD). In addition, a storage area is required for the waste.

For biogas, retention time is limited for cattle manure 20–25 days; for other animal waste 12–15 days.

Siting facilities can be difficult because of state regulations as well as public sentiment—“not in my backyard.”

3.7.13 Wind power

3.7.13.1 Characteristics

Wind turbines range in size from 1 to 100 meters in rotor diameter and from 100 watts to 5000 kilowatts in power output. Wind turbines suitable for residential- or village-scale wind power range from 500 W to 50 kW. Even though wind is an intermittent source, on a large grid it can contribute an estimated 15%–20% of annual electricity production without special arrangements for storage, backup, and load management.

The amount of energy in wind speed is proportional to the cube of the wind speed. While wind speed varies over time, it generally follows daily and seasonal patterns. Utility-scale wind power plants require wind speeds of at least 13 mph (6 m per second). A 10 kW turbine located in a moderate wind regime can generate an average of 30 kWh of power each day. For large-scale projects, 12 months of consistent observation and recording is recommended for assessing wind resources.

Wind energy produces no GHG emissions. Thus, every kWh of electricity generated by wind technologies avoids the emissions associated with a similar number of fossil-fuel-generated kWh.

Advances in the fields of aerodynamics and composite materials have helped reduce the costs of wind turbines, making modern electric power-generating wind turbines a reality. Utility-interconnected wind turbines can generate power synchronous with the grid. These machines are economically attractive where there is a good wind resource and local power costs exceed 15 cents (and sometimes less) per kilowatt hour. State-of-the-art wind technology can operate with 98% availability, and today’s turbines perform with capacity factors over 30%.

SIZE: 100–1000 kW (utility-scale); 1–50 kW (distributed power)

FEATURES: Grid-connected or stand-alone uses, but availability depends on the presence of wind. Well-designed and well-maintained wind turbines at windy sites can generate 1000 kWh/m/year.

COST: $1000–1200/kW, (utility-scale) (1992 dollars) $1900–$2200/kW, (distributed, grid-connected) $2400–$5600/kW, (distributed, battery storage). Cost is very dependent on average annual wind speed, but under ideal conditions, electricity can be generated from wind for as little as $0.04/kWh, making wind competitive with conventional fuels.

Costa Rica has installed a wind power plant of 6.4 MW that will annually save some $3.8 million in imported fuel oil costs and reduce CO, by 38,600 tons over 4 years.

CURRENT USAGE: Nearly 8000 MW worldwide at the end of 1997, although several more have been proposed.

POTENTIAL USAGE: Total worldwide wind potential is enormous; in China alone, total wind energy potential is estimated at 250,000 MW.

3.7.13.2 Issues with implementing action

There is esthetic opposition to wind because of operating noise and turbine locations. However, turbines can be located in rural areas, with surrounding land used for agriculture or other purposes.

Birds are attracted to the whirring noises made by the turbines; in some areas, bird mortality rates have increased significantly, in some instances affecting endangered species of birds.

Resources are site specific and may not be located close to demand centers.

Wind is intermittent; if not grid connected, a source of backup power is needed, increasing the costs of generation.

3.7.13.3 Climate change impact

EMISSION EFFECT: AVOIDED

3.7.13.4 Conditions for emissions mitigation

If wind potential reaches the projected 700–1000 TWh worldwide by 2020, it would avoid the production of 0.1–0.2 GtC/year of fossil fuel-fired electricity.

EMISSION ESTIMATE: 1 kWh of wind avoids 0.5–1.0 kg/CO, (-one to two Ibs/CO) from conventional sources. A wind turbine with a 500-kW capacity operating at 309% availability and producing 1.3 MWh per year avoids 351 MtC/year.

COST-EFFECTIVENESS: $21.53 ton/C

SECONDARY EFFECTS: Produces no air pollutants or GHGs, and generation avoids up to 7 g/kWh of SO, NO_x, and particulates from the coal fuel cycle (including mining and transport); 0.1 g/kWh of trace metals (including mercury); and more than 200 g/kWh of solid wastes from coal tailings and ash.

3.7.14 Recycling of coal-combustion by-products

3.7.14.1 Characteristics

Coal-fired electric power plants produce considerable solid by-products, exceeding 100 million tons per year in the United States, primarily in the form of fly ash (80%) and bottom ash, of which less than 30% is recycled for productive purposes. These amounts only increase with the rise in the number of power plants that use pollution control technologies.

Fly ash can be used in many cement and concrete applications as a substitute for portland cement. The manufacture of portland cement requires considerable amounts of energy, which generates GHG emissions, and emits CO_2 during the calcination process.

Currently, less than 25% of US fly ash is utilized in any form, and less than 15% is being used as a portland cement substitute. Gypsum and other sludges generated from the flue gas desulphurization process can replace raw gypsum in industrial and agricultural processes. The energy costs and resulting emissions associated with the acquisition and use of raw gypsum can be avoided by recycling the desulphurization process by-products.

SIZE: The capability exists for significantly increasing the use of by-products from power plants.

FEATURES: The use of by-products reduces the amount and cost of landfilling. At the same time, it could be substituted for materials, that when processed, emit GHGs.

COST: O&M costs of existing landfills are \$2–4/ton, but the Cost of developing new landfills is up to \$30/ton

CURRENT USAGE: <25% of US fly ash is utilized.

POTENTIAL USAGE: The potential exists for most of the by-products from power plants to be utilized.

3.7.14.2 Climate change impact

EMISSION EFFECT: AVOIDED

3.7.14.3 Conditions for emissions mitigation

Great potential for significant reductions in GHG emissions.

EMISSION ESTIMATE: N/A. Varies from site to site.

COST-EFFECTIVENESS: This can be very cost-effective, especially where land costs are high.

SECONDARY EFFECTS: Additional benefit derived from eliminating the need for landfills.

3.7.14.4 Issues with implementing action

Must have a sufficient market for by-products that is within economic distance of the power plant.

The tipping costs for by-products must be less than landfill costs.

The quality of by-products must meet market specifications.

3.7.15 Utilizing clean coal technology—fluidized bed combustion

3.7.15.1 Characteristics

Fluidized bed combustion (FBC) is a well-established power generation technology. In the combustor, a bed of crushed coal mixed with limestone is suspended on jets of air, tumbling in a manner that resembles a boiling liquid, hence the name "fluidized." The limestone in the mixture acts as a chemical sponge, capturing more than 90% of the sulfur before it escapes the boiler. The resulting waste is removed with the ash, in the form of benign solid easily disposed of or used in agricultural and construction applications. The lower combustion temperature needed in the process prevents the formation of 10%—80% of the nitrogen oxides typically emitted by conventional pulverized coal boilers. FBC systems also can use high-ash coals. Compared with conventional subcritical pulverized coal steam plants, FBC provides a high sulfur capture rate without degrading thermal efficiency.

There are two types of FBC systems, atmospheric (AFBC) and pressurized (PFBC), operating at pressures 6 to 16 times higher than normal atmospheric pressure. PFBC systems can achieve higher thermal efficiencies than AFBC systems. Energy in the high-pressure gases exiting the PBFC boiler can drive both a gas and a steam turbine, known as a combined cycle system. In addition, the higher thermal efficiencies of PFBC systems result in lower carbon-containing coal fuel requirements compared with conventional pulverized coal steam plants. This results in lower GHG emissions.

Currently, AFBC is commercially available in the United States; PFBC has been demonstrated but is not in widespread commercial operations. Based on operating performance, future plants are expected to have significantly reduced capital costs.

SIZE: 10—100 MW equivalent for industrial boilers, 75—350 MW for electric utility applications.

FEATURES: New baseload generation capacity or repowering of older conventional coal-fired plants. It can burn a wide variety of low-quality coals and municipal wastes. Repowering with PFBC increases plant efficiency and can raise plant capacity by 20%–25%.

COST: AFBC (200 MW): $1500–2000/kW; AFBC (repowering): $500–$1000/kW; PFBC (demonstration): $1900–$3200/kW; PFBC (commercial): $1000–$1500/kW (expected)

CURRENT USAGE: Approximately 300 AFBC units supply heat to industrial processes, municipalities, oil producers, and farms in the United States and Europe. One 70 MW PFBC demonstration project has operated in the United States with a second 145 MW, demonstration project scheduled to begin operations in 2002. A 135 MW plus 224 MW heat PFRC operates in Stockholm Sweden. Single PBFC plants are also in operation in Spain (79 MW Escatron project) and Japan.

POTENTIAL USAGE: Older conventional coal-fired plants considering life extension or retirement could be repowered using FBC technology in the United States alone, over 100 GW of capacity are already greater than 30 years old and are therefore potential candidates. Demand for new coal-fired generation capacity. Where controls on sulfur and nitrogen oxide emissions must be included, they are also candidates for FBC technology.

3.7.15.2 Climate change impact

EMISSION EFFECT: AVOIDED

3.7.15.3 Conditions for emissions mitigation

AFBC efficiencies reduce carbon emissions by approximately 3% compared with conventional steam coal plants.

Near-term PFBCs that achieve efficiencies of 40%–45% will reduce carbon emissions by 17%–27% compared with conventional steam coal plants with efficiencies of 33%. In the longer term, PBFC is expected to achieve 50% efficiency; carbon emissions will be reduced by 34% compared with conventional steam coal plants currently in operation.

EMISSION ESTIMATE: PFBC reduces C emissions by 17%–34% from current emissions; AFBC reduces C emissions by 3% from current emissions

COST-EFFECTIVENESS: The capital costs of these systems are high. However, they inherently reduce CO emissions.

SECONDARY EFFECTS: Higher efficiencies will also reduce associated air pollutant emissions from burning fossil fuels.

3.7.15.4 Issues with implementing action

PFBCs are currently being demonstrated, but are not yet commercially deployed.

These technologies have a high capital cost relative to new, natural-gas combined cycle technologies. Therefore, to be competitive, they must use low- and negative-cost fuels (e.g., wastes

3.7.16 Utilizing clean coal technology—integrated coal gasification combined cycle systems

3.7.16.1 Characteristics

The integrated coal gasification combined cycle process reacts coal with high-temperature steam and an oxidant in a reducing atmosphere to form a fuel gas The fuel gas is either passed directly to a hot-gas cleanup system to remove particulates and sulfur and nitrogen compounds or cooled to produce steam and then cleaned conventionally. The clean fuel gas is combusted in a gas turbine generator. with residual heat in the exhaust gas recovered in a heat recovery steam generator and turbine.

More than 95% of the sulfur can be removed from coal. and 90% of the nitrogen is captured. In addition, the higher thermal efficiencies of IGCC systems result in lower carbon-containing coal fuel requirements when compared with current conventional pulverized coal steam plants, resulting in lower GHG emissions.

SIZE: 200–800 MW, modular designs of 50–150 MW, may be the basis for future IGCC power plants.

FEATURES: 40% efficiency (demonstration plants) 45% efficiency (first generation commercial plant) 50% efficiency (second generation plant)

COST: $1200–$3000/kW for demonstration plants. $1200–$1900/kW for first generation plants. <$1000/kW (projected) for second-generation plants.

CURRENT USAGE: Approximately 10 demonstration plants in the United States and Europe. Several commercial plants using refinery wastes are in operation or under construction.

POTENTIAL USAGE: IGCC has the potential for new baseload generation capacity or repowering of older conventional coal-fired plants considering life extension or repowering. In the United States alone, potential candidates include over 100 GW of capacity that is more than 30 years old. IGCC also has potential where planned coal-fired capacity additions are subject to strict controls on sulfur and nitrogen oxide emissions.

3.7.16.2 Climate change impact

EMISSION EFFECT: AVOIDED

3.7.16.3 Conditions for emissions mitigation

High efficiencies (40% demonstrated and 50% expected) will reduce carbon emissions by 27%—349% compared with conventional steam coal plants.

EMISSION ESTIMATE: 27%—34% reduction from current emissions

COST-EFFECTIVENESS: High capital costs. However, where natural gas costs are high, IGCC may be competitive.

SECONDARY EFFECTS: Significant SO, and NO, reductions; production Of wastes significantly reduced; potential to safely utilize a variety of wastes, including hazardous materials.

3.7.16.4 Issues with implementing action

IGCC is still in demonstration and is not yet considered commercial when using coal.

Capital costs are currently much higher than some other options.

3.8 Nuclear reactors

As mentioned in the material resources above, nuclear technologies are having a new coming due to the lack of emissions and the advent of SMRs that come with lower production and installation costs, short construction times, and more flexibility in operation. The development of SMRs began in the 1960s with mostly military applications and is now entering the energy sectors of the economy.

Source in figure (1967).

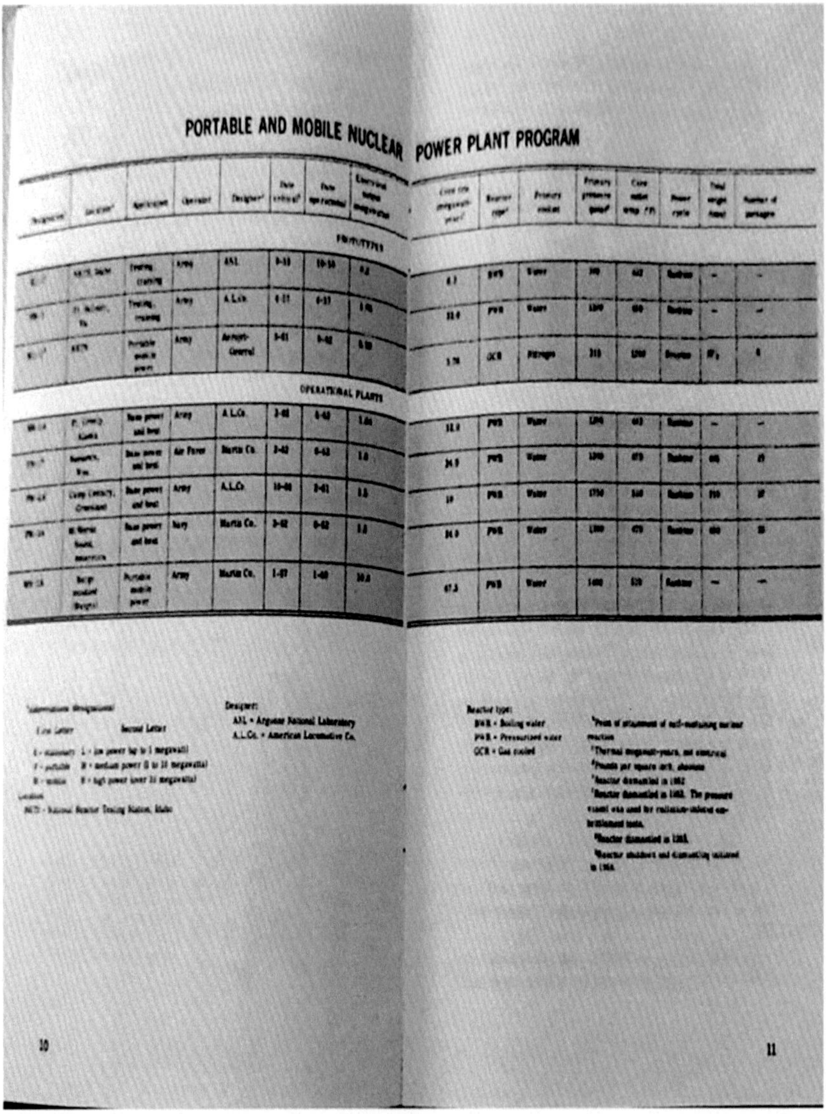

PORTABLE AND MOBILE NUCLEAR POWER PLANT PROGRAM

Designation	Location	Application	Operator	Designer	Date critical	Date operational	Electrical output megawatts	Core life megawatt-years	Reactor type	Primary coolant	Primary pressure	Core outlet temp (°F)	Power cycle	Total weight (tons)	Number of packages
PROTOTYPES															
[illegible]	NRTS, Idaho	Testing, training	Army	ANL	[illegible]	[illegible]	[illegible]	[illegible]	BWR	Water	[illegible]	[illegible]	Rankine	–	–
[illegible]	Ft. Belvoir, Va.	Testing, training	Army	A.L.Co.	[illegible]	[illegible]	[illegible]	[illegible]	PWR	Water	[illegible]	[illegible]	Rankine	–	–
[illegible]	NRTS	Portable mobile power	Army	Aerojet-General	[illegible]	[illegible]	[illegible]	[illegible]	GCR	Nitrogen	[illegible]	[illegible]	Brayton	[illegible]	[illegible]
OPERATIONAL PLANTS															
[illegible]	Ft. [illegible], Alaska	Base power and heat	Army	A.L.Co.	[illegible]	[illegible]	[illegible]	[illegible]	PWR	Water	[illegible]	[illegible]	Rankine	–	–
[illegible]	[illegible], Wyo.	Base power and heat	Air Force	Martin Co.	[illegible]	[illegible]	[illegible]	[illegible]	PWR	Water	[illegible]	[illegible]	Rankine	[illegible]	[illegible]
[illegible]	Camp Century, Greenland	Base power and heat	Army	A.L.Co.	10-60 (?)	[illegible]	[illegible]	[illegible]	PWR	Water	[illegible]	[illegible]	Rankine	[illegible]	[illegible]
[illegible]	McMurdo Sound, Antarctica (?)	Base power and heat	Navy	Martin Co.	[illegible]	[illegible]	[illegible]	[illegible]	PWR	Water	[illegible]	[illegible]	Rankine	[illegible]	[illegible]
[illegible]	Barge mounted (?)	Portable mobile power	Army	Martin Co.	1-67 (?)	[illegible]	[illegible]	67.3 (?)	PWR	Water	[illegible]	[illegible]	Rankine	–	–

Designer:
ANL = Argonne National Laboratory
A.L.Co. = American Locomotive Co.

Reactor type:
BWR = Boiling water
PWR = Pressurized water
GCR = Gas cooled

10

11

As seen in the picture above, small nuclear reactors were developed in the early 1960s and combined with nuclear reactors from the NAVY, accumulating approximately 5700 reactor years without any event. This is triggering, in the present, a frenzy of SMR and microreactor designs.

A recent survey by the Nuclear Energy Agency of the OECD lists more than 40 SMR designs. Below are only the main types, but the sky is the limit for new designs.

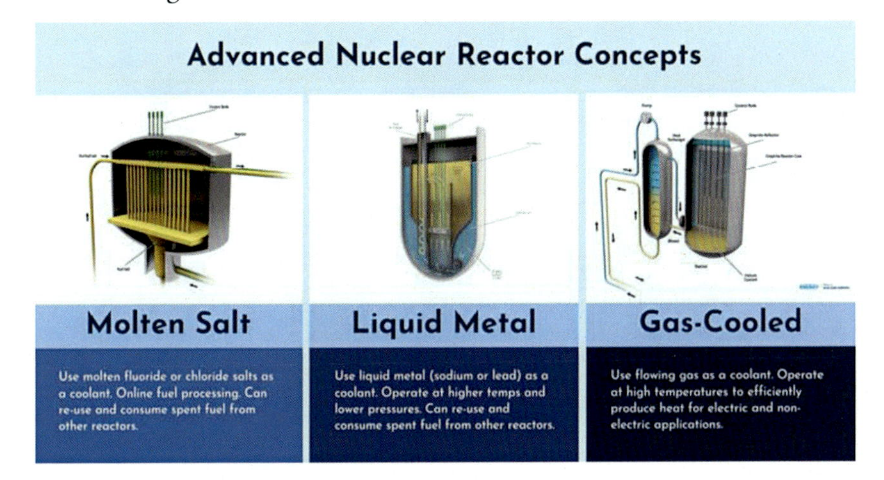

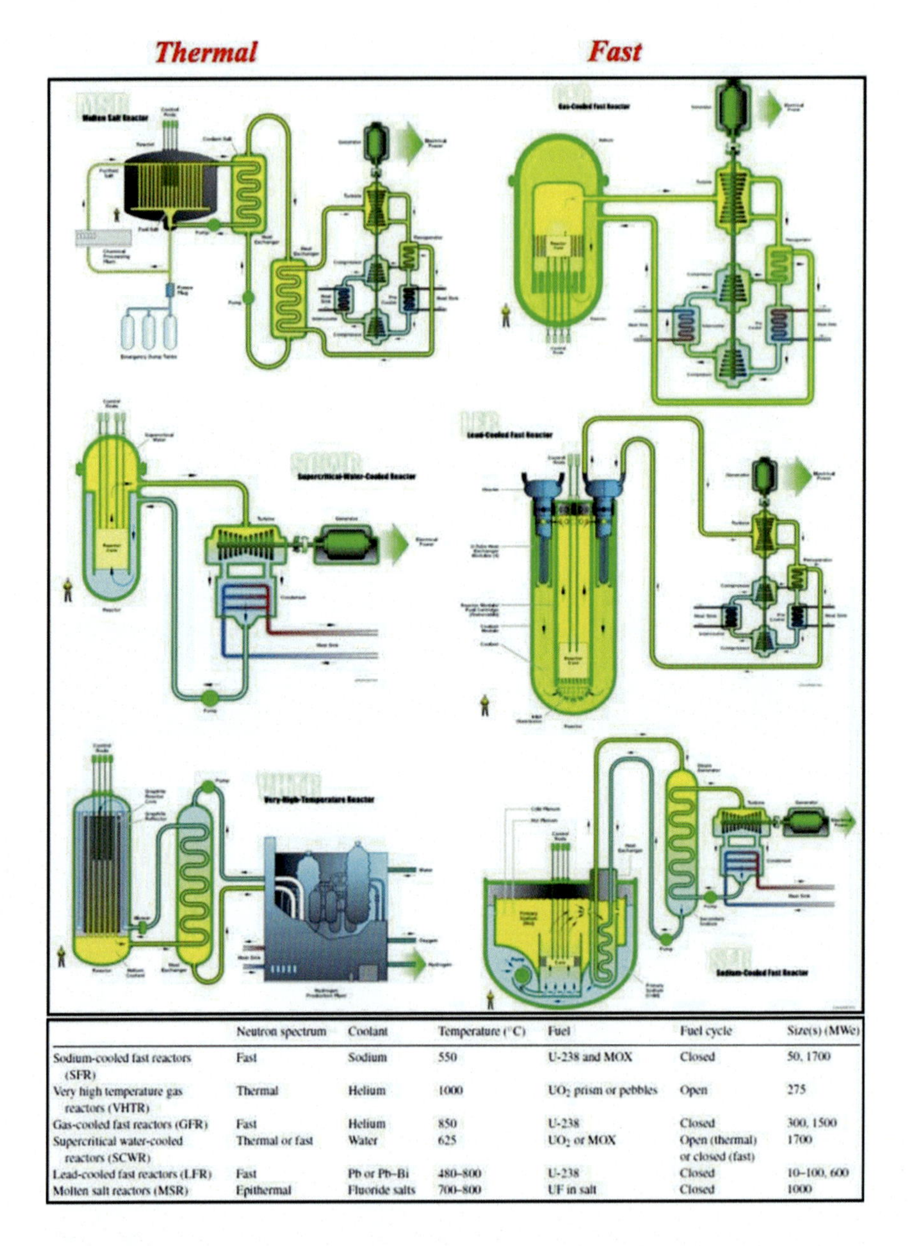

	Neutron spectrum	Coolant	Temperature (°C)	Fuel	Fuel cycle	Size(s) (MWe)
Sodium-cooled fast reactors (SFR)	Fast	Sodium	550	U-238 and MOX	Closed	50, 1700
Very high temperature gas reactors (VHTR)	Thermal	Helium	1000	UO_2 prism or pebbles	Open	275
Gas-cooled fast reactors (GFR)	Fast	Helium	850	U-238	Closed	300, 1500
Supercritical water-cooled reactors (SCWR)	Thermal or fast	Water	625	UO_2 or MOX	Open (thermal) or closed (fast)	1700
Lead-cooled fast reactors (LFR)	Fast	Pb or Pb–Bi	480–800	U-238	Closed	10–100, 600
Molten salt reactors (MSR)	Epithermal	Fluoride salts	700–800	UF in salt	Closed	1000

Regarding the sky, some nuclear propulsion designs will supposedly seriously shorten the time to reach Mars or the outer solar system's planets and moons.

Nuclear designs are now developing (again see above) microreactors for remote applications.

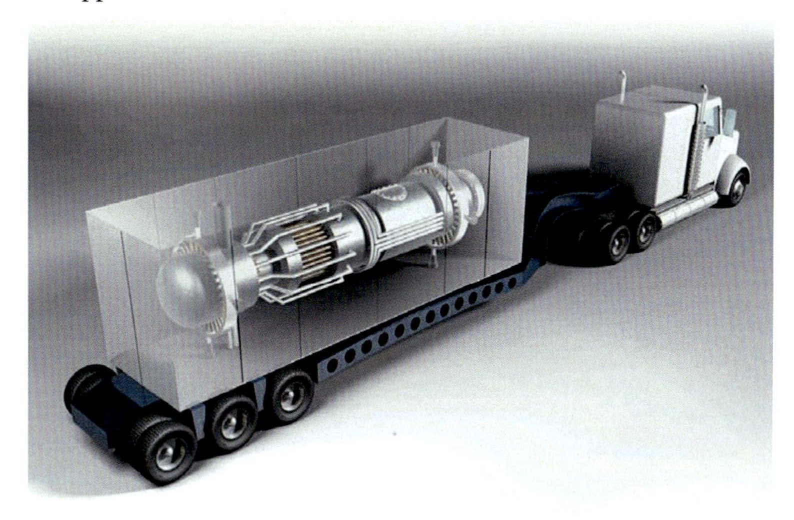

A too advanced for its time, as an example, is the Ford car with nuclear propulsion Nucleon, proposed in 1958.

The main designers are summarized in the figures below.

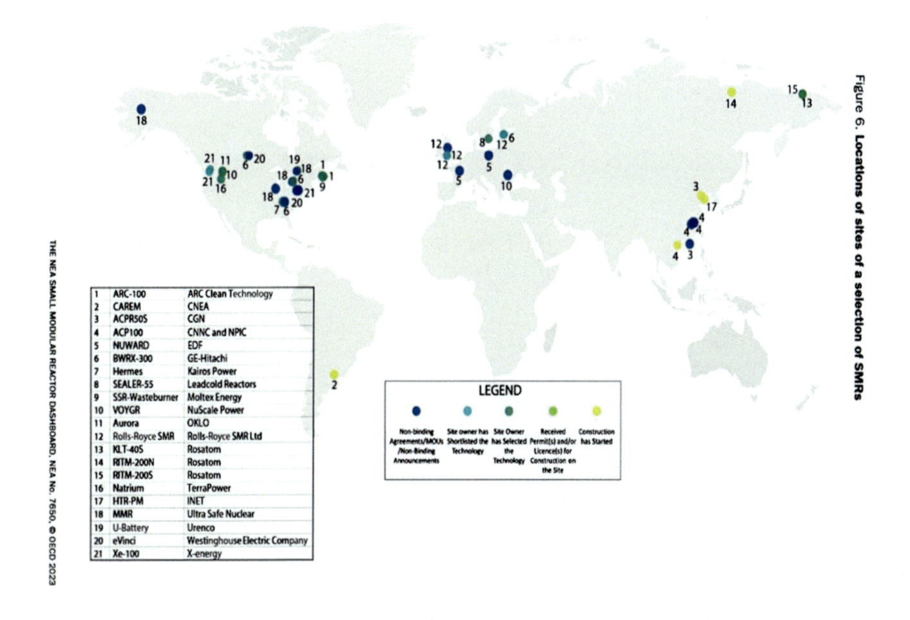

1	ARC-100	ARC Clean Technology
2	CAREM	CNEA
3	ACPR50S	CGN
4	ACP100	CNNC and NPIC
5	NUWARD	EDF
6	BWRX-300	GE-Hitachi
7	Hermes	Kairos Power
8	SEALER-55	Leadcold Reactors
9	SSR-Wasteburner	Moltex Energy
10	VOYGR	NuScale Power
11	Aurora	OKLO
12	Rolls-Royce SMR	Rolls-Royce SMR Ltd
13	KLT-40S	Rosatom
14	RITM-200N	Rosatom
15	RITM-200S	Rosatom
16	Natrium	TerraPower
17	HTR-PM	INET
18	MMR	Ultra Safe Nuclear
19	U-Battery	Urenco
20	eVinci	Westinghouse Electric Company
21	Xe-100	X-energy

Figure 6. **Locations of sites of a selection of SMRs**

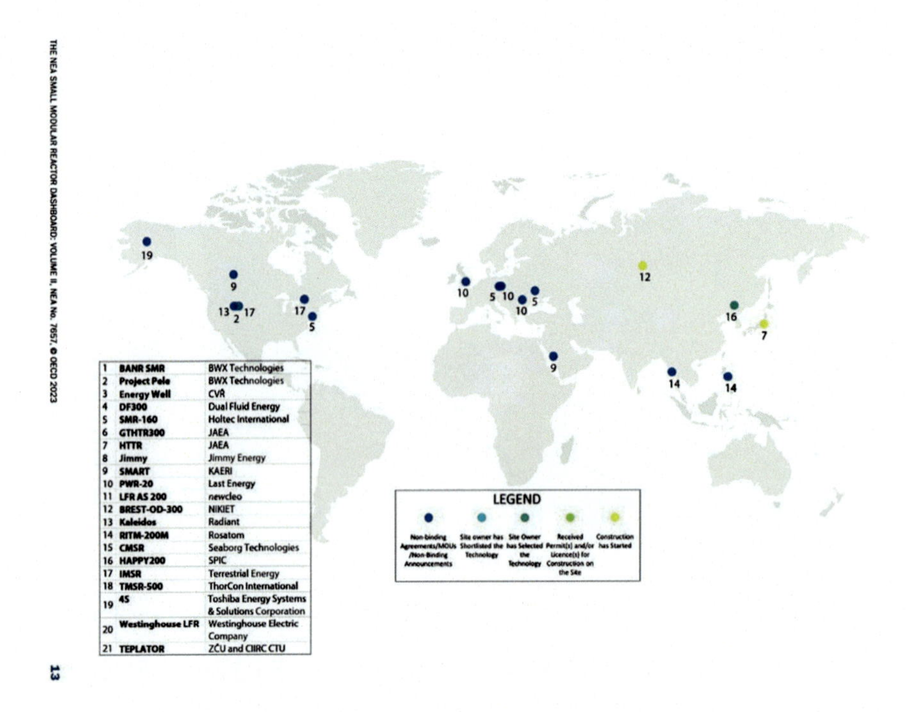

As an example of the presently most advanced, especially in terms of certification, SMR designs the figure below lists six types envisaged in the public domain for being selected in the United Kingdom. One of the designs will be implemented in the near future.

One special application is thorium-based nuclear reactors. Back in 1972 a report from Oak Ridge National Laboratory described a thorium-based reactor. Unfortunately, the work did not continue into the development of a demo reactor. Recently a thorium reactor was built in China. Since the reserves of thorium are quite large, the estimated coverage of the Chinese economy's energy needs with thorium reactors is on the order of 20,000 years. India is also exploring the technology as well as other countries (see also the Thorium Network).

The fusion power generation technologies are also having a moment of development. The old saying, from the last 60 years, that fusion is 30 years away, seems to come close to reality. Various designs (including space propulsion) are under testing. Tokamak, Stellarator, Laser fusion, as well as various more exotic ideas are now under testing with promising results. As for now, Lawrence—Livermore Laboratory in the USA has managed to obtain a regime of operation of a laser fusion installation that achieves more energy from the reaction than the one introduced, so-called *ignition*. In addition, the tokamak fusion reactor in Japan has obtained ignition preparing the operation of the ITER reactor under construction in France. These as well as other designs are the basis for the announcement of DOE that fusion will be available 10 years from now.

A new energy paradigm is being structured, and in one generation, the world's energy landscape will definitely be different.

The future of nuclear technologies is expected to bring a considerable number of applications from microreactors to be used in remote places for providing both electrical and thermal energy, to SMRs on ships or barges to supply faraway cities, to large powerplants consisting of several units of SMRs. Nuclear propulsion is going to shorten the trips to Mars or other space bodies.

Nuclear technologies are applied not only in the energy domain but also in agricultural and food industries, medical applications, construction of large objectives, industry, etc. The new nuclear uses are aimed at the production of hydrogen without emissions of CO_2, as well as district heating sources.

3.9 Energy storage technologies

The main conclusion from the rebirth of nuclear technologies is similar to the one drawn for other technologies, i.e., no resource should be neglected for evolution.

It is the creative technological environment that must be built to change liabilities into assets for each resource. The lack of such innovation and creativity may give rise to misinformation related to the specific resource that leads to losses and costs, some of which may be irreversible. In such an environment, it is important to have competent people in the decision positions aware of the real priorities for development.

As an example, the logic behind the decision to shut down coal use in energy generation based on coal being a pollutant does not consider the innovation described above. With the same logic, transport should be stopped because it is the largest polluting source locally. How would an economy function without transportation? Why neglect an important resource like coal? The optimal approach is to replace the polluting technologies in coal (with clean coal ones) or in transportation (with electric vehicles or fuel cells) such that it reduces pollution, at least locally.

Before providing an example of a potential decision about biofuel versus food in partitioning agricultural land, we present some data on energy-storage technologies, which are increasingly necessary due to the penetration of volatile renewables and electric vehicles. Additionally, a synthesis of direct conversion technologies is presented.

The EU Advisory Group for Energy has considered the potential technologies for energy storage, as presented below.

Energy storage technologies & P2X

There is no 'silver bullet' or one solution fits all.

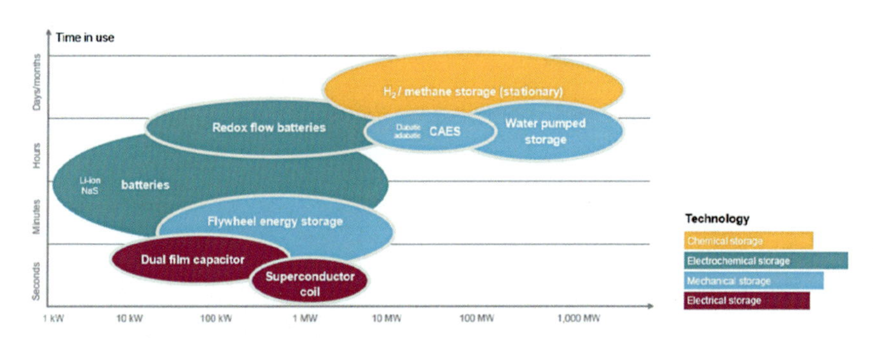

Source: Study by DNK/WEC "Energie für Deutschland 2011", Bloomberg – Energy Storage technologies Q2 2011
CAES – Compressed Air Energy Storage

Power to X

Hydrogen is the fundamental technology for Power2Chemicals

Objectives:

- Connection to 10 MW wind-farm and local Network (20 kV).
- Develop an energy storage plant in order to provide grid services (balancing mechanisms to avoid grid bottlenecks).
- Injection in local gas grid and multi-use trailer-filling.
- New conditioning concept (ionic wet gas compressor).
- Demonstrating safe handling of hydrogen and create awareness in public, politics

Technical and production aspects:

- 6 MW Electrolyzer (3 Stacks à 2 MW peak) delivered in 07/2015
- 1000 kg storage (33 MWh)
- 200 tons target annual output.

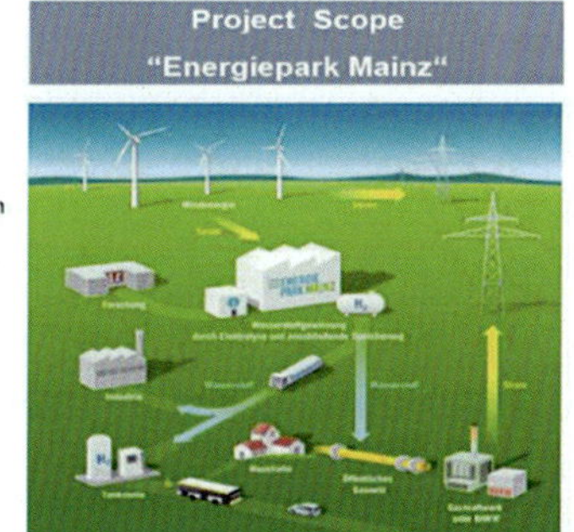

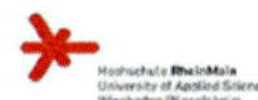

Innovation not only happens at the cell level
Battery Management and Operation in the Cloud

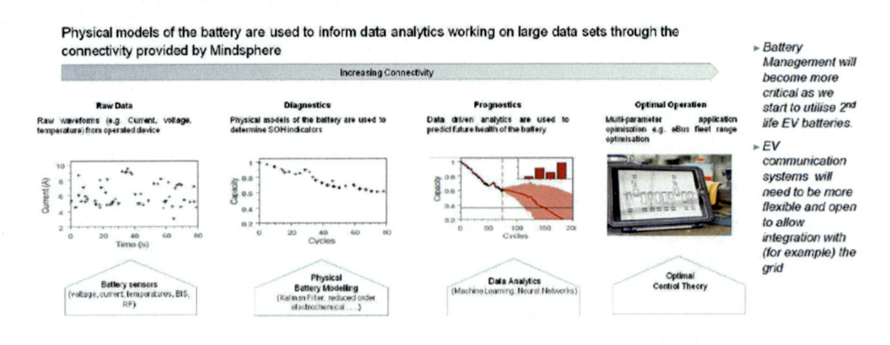

Figures courtesy of David Howey & Robert Richardson, University of Oxford

Integration of Storage & Buildings
Focus on SI, policy and Skills development

- *Within the H2020 STORY project, storage is integrated at different levels:*
  - *buildings (residential and commercial),*
  - *Micro-grids (industrial and residential)*
  - *MV distribution grids*
- Integration of storage in domestic building is still a market in development. Design of Intelligent appliances are not coordinated with storage design .
  - Battery management systems (BMS) are not communicating well
  - Limited experience in designing, building and managing the batteries
- Building level optimization becomes challenging when EV and building management systems are not integrated/coordinated
- Lack of clear regulation on the position of storage: is it generation or not?
- Lack of clearness on impact:
  - In Flanders you have to pay a fixed fee for the use of the grid in case you have PV
  - When installing batteries, a good battery management can support the grid. But there is not regulation on the assessment of the positive effect
- The combined integration of thermal and electrical storage in single buildings requires skilled people during the conceptual design. When it comes to the control .... It is nearly impossible.
  - There is a need for projects on optimized flexible control for the full management of on site generated heat and electricity with direct use, short term storage, static and dynamic storage and seasonal storage.

Besides pump storage based on hydropower plant design, energy storage technologies are diversifying at both residential and power system scales.

Another example of accelerated innovation by environmental processes is the Ozone hole. Back in the 1980s satellites discovered a growing ozone hole in the earth's atmosphere. Analysis of the process determined that the cause was substances such as chlorofluorocarbons, e.g., Freon. At the time Freon was widely used for refrigeration, sprays, air conditioning, etc. Eliminating Freon would have had a great impact on industries that depended on it for the preservation of food, with consequences for the whole planet. The Montreal Protocol in 1988 was possible because research on the topic provided positive results for replacing and eventually eliminating Freon use.

Economically this brought a new period of royalties for the companies that had developed and patented the new substances.

Related to royalties, it is important to mention the advantage in financial and economic terms for the companies that invest in innovation. Moreover, the restructuring of the power market has opened the way for new energy technologies and financial instruments. A caution is that new risks beyond those that are technical and environmental have come to the energy domain through these financial instruments.

Technologies using direct conversion are a group with the promise of wider use if fully implemented.

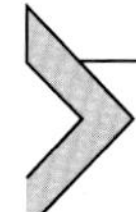

3.10 Direct conversion of energy

3.10.1 Introduction

A flashlight battery supplies electricity without moving mechanical parts. It converts the chemical energy of its contents directly into electrical energy.

Early direct conversion devices such as Volta's battery, developed in 1795, gave the scientists Ampere, Oersted, and Faraday their first experimental supplies of electricity. The lessons they learned about electrical energy and its intimate relation with magnetism spawned the mighty turboelectric energy converters-steam and hydroelectric turbines which power modern civilization.

We have improved upon Volta's batteries and have come to rely on them as portable, usually small, power sources, but only recently has the challenge of nuclear power and space exploration focused our attention on new methods of direct conversion.

To supply power for use in outer space and at remote sites on earth, we need reliable power sources, light in weight, and capable of unattended operation for long periods. Nuclear power plants using direct conversion techniques hold the promise of surpassing conventional power sources in these attributes. In addition, the inherently silent operation of direct conversion power plants is an important advantage for many military applications. The Atomic Energy Commission, the Department of Defense, and the National Aeronautics and Space Administration collectively sponsor tens of millions of dollars worth of research and development in the area of direct conversion each year. In particular, the Atomic Energy Commission supports more than a dozen research and development programs in thermoelectric and thermionic energy conversion in industry and at the Los Alamos

Scientific Laboratory and other direct conversion research at Argonne National Laboratory and Brookhaven National Laboratory. Reactor and radioisotopic power plants utilizing direct conversion are being produced under the AEC's SNAP program. Some of these units are presently in use powering satellites, Arctic and Antarctic weather stations, and navigational buoys.

Further applications are now being studied, but the cost of direct conversion appears too great to permit its general use for electric power in the near future. Direct techniques will be used first where their special advantages outweigh the higher cost.

3.10.2 Direct versus dynamic energy conversion

3.10.2.1 Dominance of dynamic conversion

We live in a world of motion. A primary task of an engineer is to find better and more efficient ways of transforming energy locked in the sun's rays or fuels, such as coal and the uranium nucleus, into energy of motion. Almost all the world's energy is now transformed by rotating or reciprocating machines. We couple the energy of exploding gasoline and air with our automobile's wheels with a reciprocating engine. The turbogenerator at a hydroelectric plant extracts energy from falling water and turns it into electricity. Such rotating or reciprocating machines are called dynamic converters.

A New Level of Sophistication: Direct Conversion

A revolution is in the making. We now know that we can force the heat-and-electricity-carrying electrons residing in matter to do our bidding without the use of shafts and pistons. This is a leading accomplishment of modern technology—energy transformation without moving parts. It is called direct conversion.

Direct conversion devices convert heat into electricity without moving parts.

3.10.2.2 Why is direct conversion desirable?

There are places where energy conversion equipment must run for years without maintenance or breakdown, Additionally, there are situations where the ultimate in reliability is required, such as on scientific satellites and particularly on manned space flights. Direct conversion equipment seems to offer greater reliability than dynamic conversion equipment for these purposes.

We should recognize that our belief in the superiority of direct conversion is based more on intuition than proof. Indeed, direct converters will never throw piston rods or run out of lubricant. Yet, some satellite power failures have been caused by the degradation of solar cells under the bombardment of solar protons. The other types of direct conversion devices described in the following pages may also break down in ways as yet unknown. Still, today's knowledge gives us hope that direct conversion will be more reliable and trustworthy than dynamic conversion. Direct conversion equipment is beginning to be adopted for small power plants, producing less than 500 W, designed to operate for long periods in outer space and under the ocean. Some day, large central-station power plants may use direct conversion to improve their efficiency and reliability.

3.10.2.3 Laws governing energy conversion

The Big Picture: Thermodynamics

To the best of our knowledge, energy and mass are always conserved together in any transformation. This law applies to everything we do, from driving a nail to launching a space probe.

In direct conversion processes we need not worry about these mass changes, but at each point we must make sure that all energy is accounted for. For example, in outer space, all energy released from fuels (even food) must ultimately be radiated away to empty space. Otherwise, the vehicle temperature will keep rising until the spaceship melts.

Direct conversion devices are no exception. Consequently, every thermoelectric element or thermionic converter must provide for the disposition of waste heat. The designer will try, however, to make the engine efficiency high enough that the waste heat will be small.

Some space power plants contemplate using space cabin heat (T 300 K) to drive a heat engine that rejects its waste heat to liquid-hydrogen rocket fuel stored at Te 20 K. What would be the Carnot efficiency of this engine?

3.10.2.4 Thermoelectricity

After 140 Years: Seebeck Makes Good

The oldest direct conversion heat engine is the thermocouple. Take two different materials (typically, two dis-similar metal wires), join them, and heat the junction. A voltage, or electromotive force, can be measured across the unheated terminals. T. J. Seebeck first noticed this effect in 1821 in his laboratory in Berlin, but, because of a mistaken interpretation of what was involved, he did not seek any practical application for it. Only recently has any real progress been made in using his discovery for power production.

To use the analogy of A. F. Joffe, the Russian pioneer in this field, thermoelectricity lay undisturbed for more than 100 years like Sleeping Beauty. The Prince that awoke her was the semiconductor.

As long as inefficient metal wires were used, textbook writers were correct in asserting that thermoelectricity could never be used for power production. The secret of practical thermoelectricity is therefore the creation of better thermoelectric materials (Creation is the right word since the best materials for the purpose do not exist in nature.) To perform this alchemy, we first have to understand the Seebeck effect.

3.10.2.4.1 Electrons and holes

Let us examine the latticework of atoms that make up any solid material. In electrical insulators, all the atoms' outer electrons are held tightly by valence bonds to neighboring atoms. In contrast, any metal has many relatively loose electrons that can wander freely through its latticework. This is what makes metals good conductors.

A thermoelectric couple is made from p- and n-type semiconductors. The impurity atoms are different in each part and contribute an excess or a deficiency of valence electrons. Heat drives both holes and electrons toward the cold junction.

The electron-hole model does not have the precision the physicist likes, but it helps us to visualize semiconductor behavior.

The Seebeck effect is demonstrated when pieces of p- and n-type materials are joined. Heat at the hot junction drives the loose electrons and holes toward the cold junction. Think of the holes and electrons as gases being driven through the latticework by the temperature difference. A positive and a negative terminal are thus produced, providing a source of power. The larger the temperature difference, the larger the voltage difference. Note that just one thermocouple leg can produce a voltage across its length, but couples made from p and n legs are superior.

3.10.2.4.2 Practical thermoelectric power generators

The first nuclear-heated thermoelectric generator was built in 1954 by the Atomic Energy Commission's Mound Laboratory in Miamisburg, Ohio. It used metal-wire thermocouples. In contrast, better thermocouples are made of thick lead telluride (PbTe) semiconductor cylinders about 2 inches long. In contrast to the thermocouple wires' efficiency of less than 1%, these last generators have overall efficiencies exceeding 5%. This value is still low compared with the 35%–40% obtained in a modern steam power plant, but these generators can operate unattended in remote locations where steam plants would be completely unacceptable.

Underlying the apparent simplicity of the thermoelectric generator are extensive development efforts, requiring thousands of experimental brazing tests. It turns out to be uncommonly difficult to fasten thermo-electric elements to the so-called hot shoe (metal plate) at the bottom. The joint has to be strong, must withstand high temperatures, and must have low electrical resistance. In addition, the elements are encased in mica sleeves to prevent chemical disturbance of the delicate balance of impurities in the semiconductor by surrounding gases. A further complication is the extreme fragility of the elements, and this is still under research.

Nuclear thermoelectric generators that provide small amounts of electrical power have already been launched into space aboard Department of Defense satellites, installed on land stations in both polar regions, and placed under the ocean.

Commercially available thermoelectric generators using propane fuel can provide more than enough electrical power to operate a portable TV set. The Russians have long manufactured a kerosene lamp with thermoelements placed within its stack for generating power in wilderness areas. Some recent applications combine solar panels with thermoelectric generators to generate some power during nighttime when temperature differences allow it.

For the present, the role of thermoelectric power appears to be one of special uses such as those just mentioned. When higher efficiencies are attained, thermoelectric power may one day supplant dynamic conversion equipment in certain low-power applications regardless of location.

3.10.2.5 Thermionic conversion

3.10.2.5.1 "Boiling" electrons out of metals

Like the thermoelectric element, the thermionic converter is a heat engine. In its simplest form it consists of two closely spaced metallic plates and resembles the diode radio tube. Whereas thermoelectric elements depend on heat to drive electrons and holes through semiconductors to an external electricity-using device or load, the salient feature of the thermionic diode is thermionic emission, or simply the boiling off of electrons from a hot metal surface.

Metals, as we have already seen, have an abundance of loosely bound conduction electrons roaming the atomic latticework. These electrons are easily moved by electric fields while within the metal, but it takes considerably more energy to boil them out of the metal into free space. Work must be done against the electric fields set up by the surface layer of atoms, which have unattached valence bonds on the side facing empty space. The energy required to completely detach an electron from the surface is called the metal's work function. In the case of tungsten, for example, the work function is about 4.5 electron volts of energy. As we raise the temperature of a metal, the conduction electrons in the metal also become hotter and move with greater velocity. We may think of some electrons in a metal as forming a kind of electron gas. Some electrons will gain such high speeds that they can escape the metal surface.

This happens when their kinetic energy exceeds the metal's work function.

Now that we have found a way to force electrons out of the metal, we would like to make them do useful electrical work. To do this we have to push the electrons across the gap between the plates as well as create a voltage difference to go with the hoped-for current flow.

3.10.2.5.2 Reducing the space charge

The emitted or boiled-off electrons between the converter plates form a cloud of negative charges that will repel subsequently emitted electrons back to the emitter plate unless counteraction is taken. To circumvent these space charge effects, we fill the space between the plates with a gas containing positively charged particles. These mix with the electrons and neutralize their charge. The mixture of positively and negatively charged particles forms a plasma. The presence of the plasma makes the gas a good conductor. The emitted electrons can now move easily across it to the collector where, continuing with the gas analogy, they condense on the cooler surface.

Thermionic converters may be flat-plate or cylindrical types. The cylindrical converter is an experimental type for ultimate use in nuclear reactors.

Result: A Plasma Thermocouple

Unless a voltage difference exists across the plates, no external work can be done. In the thermocouple, the voltage difference was caused by the different electrical properties of the p and semiconductors. Both the emitter and the collector in the thermionic converter are good metallic conductors rather than semiconductors, so a different tack must be taken.

The key is the use of an emitter and a collector with different work functions. If it takes 4.5 electron volts to force an electron from a tungsten surface and if it regains only 3.5 electron volts when it condenses on a collector with a lower work function, then a voltage drop of 1 V exists between the emitter and collector.

To summarize, then, the thermionic emission of electrons creates the potentiality of current flow. The difference in work functions makes the thermionic converter a power producer.

There is an interesting comparison that helps describe this phenomenon. Consider the emitter to be the ocean surface and the collector a mountain lake. The atmospheric heat engine vaporizes ocean water and carries it to cooler mountain elevations where it condenses as rain that collects in lakes. The lake water, as it runs back toward sea level, can then be made to drive a hydroelectric plant with the gravitational energy it has gained in transit. The thermionic converter is similar in behavior, with a hot emitter (corresponding to the sun-heated ocean), a cooler collector (lake), electron gas (water), and different electrical voltages (gravity). Without gravity, the river would not flow, and the production of electricity would be impossible.

3.10.2.5.3 Thermionic power in outer space

Thermionic converters for use in outer space may be heated by the sun, by decaying radioisotopes, or by a fission reactor. Thermionic converters can also be made into concentric cylindrical shells and wrapped around the uranium fuel elements in nuclear reactors. The waste heat in this case would be carried out of the reactor to a separate radiator by a stream of liquid metal.

In outer space, waste heat must be radiated away. The rate at which heat is radiated is proportional to the fourth power of Te (Stefan–Boltzmann law).

Since thermionic converters can operate at much higher temperatures than thermoelectric couples or dynamic power plants, the radiator temperature, T_C, will also be higher. Consequently, space power plants using thermionic converters will have small radiators. Once thermionic converters are developed with high reliability and long life, they will provide the basis for a new series of lighter, more efficient space power plants.

3.10.2.6 Magnetohydrodynamic conversion

3.10.2.6.1 Big word, simple concept

Magnetohydrodynamic (MHD) conversion is very unlike thermoelectric or thermionic conversion. MHD generators use high-velocity electrically conducting gases to produce power and are generically closer to dynamic conversion concepts. The only concept they carry forward from the preceding conversion ideas is that of plasma, the electrically conducting gas. Yet they are commonly classified as direct because they replace the rotating turbogenerator of the dynamic systems with a stationary pipe or duct.

In a typical design, in the MHD duct, the electrons in hot plasma move to the right under the influence of force F in a magnetic field B. The electrons collected by the right-hand side of the duct are carried to the load. In a wire in the armature of a conventional generator, the electrons are forced to the right by the magnetic field.

In the conventional dynamic generator, an electromotive force is created in a wire that cuts through magnetic lines of force. It may be helpful to visualize the conduction electrons as leaving one end of the wire and moving to the other under the influence of the magnetic field.

The surge of electrons along the length of the wire sets up a voltage difference across the ends of the wire. A generator uses this difference to convert the kinetic energy of the moving wire or armature into electrical energy. The wire is kept spinning by the shaft which is connected to a turbine driven by steam or water.

Let us try to eliminate the moving part, the generator armature. What we need is a moving conductor that has no shaft, no bearings, and no wearing parts. The substance that meets these requirements is the plasma. The MHD generator substitutes a moving, conducting gas for the wires. Under the influence of an external magnetic field, the conduction electrons move through the plasma to one side of the duct which carries electrical power away to the load.

The MHD generator gets its energy from an expanding, hot gas; but, unlike the turbogenerator, the heat engine and generator are united in the static duct. The gradual widening of the duct reflects the lower pressure, cooler plasma at the duct's end. Some of the plasma's thermal energy content has been tapped off by the duct's electrodes as electrical power.

3.10.2.6.2 The fourth state of matter

Plasma can be created by temperatures over 2000 K. At this temperature, many high-velocity gas atoms collide with enough energy to knock electrons off each other and thus become ionized. The material thus created, seen as a glowing gas, does not behave consistently as any of the three familiar states of matter: solid, liquid, or gas. Plasma has been called a fourth state of matter. Since we have difficulty in containing such high temperatures on earth, we adopt the strategy of seeding. In this technique, gases that are ordinarily difficult to ionize, like helium, are made conducting by adding a fraction of a percent of an alkali metal such as potassium. Alkali metal atoms have loosely bound outer electrons and quickly become ionized at temperatures well below 2000 K.

A helium-potassium mixture is a good enough conductor for use in an MHD generator. In this plasma, the electrons move rapidly under the influence of the applied fields, though not as well as in metals. The positive ions move in the opposite direction from the electrons, but the electrons are much lighter and move thousands of times faster thus carrying the bulk of the electrical current.

3.10.2.6.3 Magnetohydrodynamic power prospects

The MHD duct is not a complete power plant in itself because, after leaving the duct, the stream of gas must be compressed, heated, and returned to the duct. Very high-temperature materials and components must be developed for this kind of service. Moreover, while the duct is conceptually simple, it must operate at very high temperatures in the presence of corrosive alkali metals. This presents us with a difficult material problem. A large part of the problem is solved, and thus, MHD power plants should be able to provide reliable power with high efficiency. They may then serve in large space power plants and submarine propulsion devices, and most importantly, they may provide cheaper electricity for general use through their higher temperatures and greater efficiencies.

3.10.2.7 Chemical batteries

3.10.2.7.1 Electricity from the chemical bond

If you vigorously knead a lemon to free the juices and then stick a strip of zinc in one end and a copper strip in the other, you can measure the voltage across the strips. Electrons will flow through the load without the inconvenience of having to supply heat. You have made yourself a chemical battery.

The chemical battery was the first direct conversion device. Two hundred years ago it was the scientists' only continuous source of electricity.

Since the chemical battery does not need heat for its operation, it is logical to ask what makes the current flow. Where does the energy come from?

The battery has no semiconductors, but, like the thermo-electric couple and the thermionic diode, it uses dissimilar materials for its electrodes. A conducting fluid or solid is also present to provide for the passage of current between the electrodes. In the example of the lemon, the copper and zinc are the dissimilar electrodes, and the lemon juice is the conducting fluid or electrolyte that supplies positive and negative ions. The battery derives its energy from its complement of chemical fuel. The voltage difference arises because of the different strengths of the chemical bonds.

The chemical bond is an electrostatic one; some atoms have stronger electrical affinities than others.

3.10.2.7.2 Chemical reactions used in batteries and fuel cells

In principle, all these reactions are the same as those going on inside the lemon, although each type of cell produces a slightly different voltage because of the varying chemical affinities of the atoms and molecules involved. Hundreds of materials can be used for electrolytes and electrodes.

No heat needs to be added as the electrostatic chemical bonds are broken and remade in a battery to generate electrical power. The chemical reaction energy is transferred to the electrical load with almost 100% efficiency. The Carnot cycle is no limitation here; only "cold" electrostatic forces are in action. The reactions cannot go on forever, however, because the battery supplies the energy converter with a very limited supply of fuel. Eventually, the fuel is consumed and the voltage drops to zero. This deficiency is remedied by the fuel cell in which fuel is supplied continuously.

3.10.2.7.3 An old standby in outer space

Almost every satellite and space vehicle has a chemical battery aboard. It is not there so much for continuous power production but as a rechargeable electrical accumulator or reservoir to provide electricity during peak loads. The battery is also needed to store energy for use during periods when solar cells are in the earth's shadow and therefore inoperative. In this capacity, the dependable old battery serves most modern science very well indeed.

3.10.2.8 The fuel cell

3.10.2.8.1 A continuously fueled battery

Potential fuels The battery has a very close relative, the fuel cell. Unlike the battery, the fuel cell has a continuous supply of fuel. The hydrogen-oxygen cell is typical of all fuel cells. It essentially burns hydrogen and oxygen to form water. If the hydrogen and oxygen can be supplied continuously and the excess water drained off, we can greatly extend the life of the battery. The fuel cell accomplishes this. Fueled electrical cells would be more descriptive since the physical principles are identical to those of a battery.

The chemical battery works in the same way except that the chemicals are different and are not continuously supplied from outside the cell. The water produced by the H—O cell shown can be used for drinking on spaceships.

Perhaps the most challenging task contemplated for the fuel cell is to bring about the consumption of raw or slightly processed coal, gas, and oil fuels with atmospheric oxygen. If fuel cells can be made to use these abundant fuels, then the high natural conversion efficiency of the fuel cells will make them economically superior to the lower-efficiency steam-electric plants now in commercial service.

So far, we have dwelt on the fuel cell as a cold energy conversion device that is not limited by Carnot efficiency. A variation on this theme is possible. Take a hydrogen iodide (HI) cell, and heat the HI to 2000 K. Some of the HI molecules will collide at high velocities and dissociate into hydrogen and iodine: 2HI H_2+ 1; the higher the temperature, the more the dissociation. By separating hydrogen and iodine gases and returning them for recycling to the fuel cell for recombination, we have eliminated the fuel supply problem and created a regenerative fuel cell. We have, however, reintroduced the heat engine and Carnot cycle efficiency. The thermally regenerative fuel cell is a true heat engine using a dissociating gas as the working fluid.

Scheme for Project Apollo Most of the impetus for developing the fuel cell as a practical device comes from the space program. The cell has admirable properties for space missions that are less than a few months in duration. It is a clean, quiet, vibrationless source of energy. Like the battery, it has a high electrical overload capacity for supplying power peaks and is easily controlled. It can even provide potable water for a crew if the Bacon H—O cell is used. For short missions where large fuel supplies are not needed, it is also among the lightest power plants available.

These compelling advantages have led the National Aeronautics and Space Administration to choose the fuel cell for some of the first space ventures with crews. Project Apollo, a staffed lunar landing mission, is the most notable example. Here the fuel cells were not only an energy source but also a part of the ecological cycle that kept the crew alive.

3.10.2.9 Solar cells

3.10.2.9.1 Photons as energy carriers

All our fossil fuels, such as coal and oil, owe their existence to the solar energy stream that has engulfed the earth for billions of years. The power in this stream amounts to about 1400 W per square meter at earth, nearly enough to continuously supply energy for an average home if all the energy were converted to electricity. The problem is getting the sun's rays to yield their energy with high efficiency.

The sun's visible surface has a temperature of around 6000 K. Any object heated to this temperature will radiate visible light mostly in the yellow-green portion of the spectrum (5500 A*). Our energy conversion device should be tuned to this wavelength.

3.10.2.9.2 Harnessing the sun's energy

Historically, the sun's energy has most often been used by concentrating it with a lens or mirror and then converting it to heat. We could do this and beat angle, but a more direct avenue is open. About a decade ago it was found that the junction between p and n semiconductors would generate electricity if illuminated. This discovery led to the development of the solar cell, a thin, lopsided sandwich of silicon semiconductors.

For example, the solar cell is in use on satellites. A spherical, radioisotope, thermoelectric generator at the bottom of the satellite is used to supplement the solar cells. In the solar cell, hole-electron pairs are created by solar photons in the vicinity of a p-junction.

Whenever p- and n-type semiconductors are sandwiched together a voltage difference is created across the junction. The separated holes and electrons in the two semiconductor regions establish this electric field across the junction. Unfortunately, there are usually no current carriers near the junction, so no power is produced.

The absorption of solar photons in the vicinity of the junction will create current carriers, as the photons' energy is transformed into the potential energy of the hole-electron pairs. These pairs would quickly recombine and give up their newly acquired potential energy if the electric field existing across the junction did not whisk them away to an external load.

The solar cell produces electricity when hole-electron pairs are formed. Any other phenomenon that creates such pairs will also generate electricity. The source of energy is irrelevant so long as the current carriers are formed near the junction. Thus, particles emitted by radioactive atoms can also produce electricity from solar cells, although too much bombardment by such particles can damage the cell's atomic structure and reduce its output.

The solar cell is not a heat engine. Yet it loses enough energy so that the sun's energy is converted at less than 15% efficiency. Losses commonly occur because of the recombination of the hole-electron pairs before they can produce current, the absorption of photons too far from the junction, and the reflection of incident photons from the top surface of the cell. Despite these losses, solar cells are now the mainstay of nonpropulsive space power.

3.10.2.10 *Nuclear batteries*

3.10.2.10.1 Energy from nuclear particles

As we have seen, solar cells can convert the kinetic energy of charged nuclear particles directly into electricity, but a more straightforward way of doing this exists. This involves direct use of the flow of charged particles as current.

The nuclear battery performs this trick. A central rod is coated with an electron-emitting radioisotope (a beta-emitter; say, strontium-90). The high-velocity electrons emitted by the radioisotope cross the gap between the cylinders and are collected by a simple metallic sleeve and sent to the load. Simple, but why do space-charge effects not prevent electrons from crossing the gap as they do in the thermionic converter? The answer is that nuclear electrons have a million times more kinetic energy than those boiled off the thermionic converter's emitter surface. Consequently, they are too powerful to be stopped by any space charge within the narrow gap.

The nuclear battery depends on the emission of charged particles from a surface coated with a radioisotope. The particles are collected on another surface.

Nuclear batteries are simple and rugged. They generate only microamperes of current at 10,000 to 100,000 V.

3.10.2.10.2 Double conversion

In the description of the energy conversion, we saw that we could go through the energy transformation process repeatedly until we obtained the kind of energy we wanted. This is exemplified in a type of nuclear battery that uses the so-called double conversion approach. First, the high-velocity nuclear particles are absorbed in a phosphor which emits visible light. The photons thus produced are then absorbed in a group of strategically placed solar cells, which deliver electrical power to the load. Although efficiency is lost at each energy transformation, the double conversion technique still ends up with an overall efficiency of from 1% to 5%, an acceptable value for power supplies in the watt and milliwatt ranges.

3.10.2.11 *Advanced concepts*

Ferroelectric and thermomagnetic conversion are subtle concepts that depend on the growth alteration of a material's physical properties by the application of heat. Devices employing such concepts are true heat engines. Instead of the gaseous and electronic working fluids used in the other direct conversion concepts, the ferroelectric and thermomagnetic concepts employ patterns of atoms and molecules that are rearranged periodically by heat.

3.10.2.11.1 Ferroelectric conversion

Ferroelectric conversion makes use of the peculiar properties of dielectric materials. Barium titanate, for example, has good dielectric properties at low temperatures, but, when its temperature is raised to more than 120°C, the properties get worse rapidly. We cannot discuss dielectric behavior thoroughly here; suffice it to say that in this process heat is absorbed in a realignment of molecules within the barium titanate latticework.

If we now place a slab of barium titanate between the two plates of an electrical condenser and charge the condenser, we have a unique way of converting heat into electricity directly. When the barium titanate is heated above its Curie point of 120°C, the condenser's capacitance is radically reduced as the dielectric constant falls. The condenser is forced to discharge and move electrons through an external circuit consisting of the load and the source of charge. Useful electrical energy is delivered during this step. When the dielectric is cooled, waste heat is given up by the barium titanate, and the cycle is complete.

Dielectric materials are nonconductors such as those used between the plates of a condenser to increase its electrical capacity. The Curie point is the temperature at which a material's crystalline structure radically changes and becomes less orderly.

3.10.2.11.2 Thermomagnetic conversion

The analog of ferroelectricity is ferromagnetism. A converter employing similar principles to those in ferroelectricity can be made using an electrical inductance with a ferromagnetic core. When the temperature of the ferromagnetic material is raised above its Curie point, its magnetic permeability drops quickly, causing the magnetic field to collapse partially. Energy may be delivered to an external load during this change. Instead of energy being stored in an electrostatic field, it is stored in a magnetic field.

*Ferroelectricity and ferromagnetism are very similar. The equations describing these phenomena are almost identical except that capacitance is replaced by its magnetic analog, inductance, and so on.

Ferroelectric and thermomagnetic conversion both represent a class of energy transformations that involve internal molecular or crystalline rearrangements of solids. There is no change of phase as in a steam engine, but the energy changes are there nevertheless. In thermodynamics, such internal geometrical changes are called second-order transitions, as opposed to the first-order transitions observed with heat engines using two-phase working fluids like water/steam.

3.10.2.11.3 On the frontier

Other potential energy conversion schemes are being investigated by scientists and engineers.

In particular, we are just learning how to manipulate photons. There are photochemical, photoelectric, and even photomechanical transformations. These have hardly been tapped.

Consider the reaction when an electron and its antimatter equivalent, the positron, meet. They mutually annihilate each other in a burst of energy! This energy will be harnessed someday. Various quantum-level processes are being researched.

What energy conversion device are we going to use to completely convert mass into energy? The energy requirements for interstellar exploration are so great that these voyages will be impossible unless a new device is found that can completely transform mass into energy.

Then again, we do not have the faintest idea of how to control gravitational energy, but we may learn.

The panorama is endless.

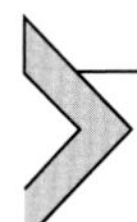

3.11 Geostrategy of resources and critical infrastructures

Most of the time, critical resources and infrastructures have determined the behavior of states and economic structures both locally and internationally. To the extent that states have evolved technologically, they have often developed their military but also economic access to resources and created critical infrastructures from the point of view of accessing resources. The types of resources considered have changed over time, but the behavior regarding their control and use has remained with a certain level of invariability.

Some examples are relevant:

Cuba has large deposits of zinc. As long as Zinc is exported, the political regime should remain stable (regardless of the type of regime).

Vietnam has deposits of uranium and titanium. As long as the benefit of what was extracted was greater than the cost of war, the war continued. When low-cost extraction was exhausted, the war stopped. Now another resource is used, i.e., cheap labor obtained through commercial networks.

Kosovo has chrome deposits, and the chrome rolling mill is in Albania. It is therefore right that there should be an independent state that can export chrome.

Afghanistan, in addition to its sovereign deposits, is also an access route for transporting gas from Uzbekistan to ports that go beyond the Strait of Hormuz, possibly controlled by Iran. If the Taliban regime accepts the transfer, then it is no longer necessary to maintain troops in Afghanistan. This could be a partial explanation for the recent withdrawal of US troops, one other explanation being that the United States had become a gas exporter and no longer needed the pipelines.

Norway has important gas deposits in the North Sea, a situation that keeps it outside the EU to maintain control over these deposits. When those deposits are exhausted, the accession of Norway to the EU could be seen.

The USA has become an exporter of oil and liquefied natural gas thanks to the extraction technologies developed in the last 15 years. Before this development, the USA was the largest importer of petroleum products.

The Russian Federation was, until the invasion of Ukraine, the largest supplier of natural gas to the EU. Through this, the EU economy is controlled so that any action hostile to Russian politics is sanctioned through the gas pipeline network (see, for example, the reaction to EU opposition to the Russian actions in Abkhazia in 2008, leading to the interruption of gas supplies for a month). Russia's advantage is that it also delivers gas to China, so a decrease in quantities to the EU can be compensated. Moreover, recently two components are controlled, quantities and prices controlled by the mechanisms of the energy market, which is used destructively for the economies of the EU. In this way, additional money is obtained at a time when it is needed to finance Russia's military measures against Ukraine and, indirectly, NATO. Paradoxically, NATO member countries in the EU were financing measures against NATO. The emergence of gas exports from the USA, as well as other sources (mainly Qatar), may diminish Russia's influence. A potential reaction to diminish the supply of LNG from Qatar with an impact on the EU is the recent attack of Hamas on Israel and the strong response that may trigger a reaction by Qatar to stop deliveries of LNG, especially at the beginning of winter. Another reason could be to increase the risks of an alternative route through India-Middle East to Europe that threatens the present one passing through the Russian Federation.

Saudi Arabia, Qatar, and the UAE are still important in the export of oil and liquefied gas, but they will gradually lose their importance due to the introduction of electric transport technologies and electricity generation based on renewables and fission or fusion in the foreseeable future. Without oil money, these countries will have to reinvent themselves, a process that is underway through investments in both tourism and nuclear power plants.

Peru and other countries with lithium reserves can become important because of Li-ion battery technology, which is needed in large quantities for electric vehicles and energy storage. Other types of batteries have been recently developed, e.g., with sulfur, which will increase the importance of the various states possessing such resources.

China has important deposits of rare earths needed for communication technologies as well as in other fields. Iraq also has such deposits and other necessary resources. It is therefore important that the economic situation is stable so that exports are made without interruption.

Mali, in Africa, is characterized by uranium ore reserves exported to the French economy based mostly on electricity from nuclear power plants. That is why possible political instability in this country triggers intervention by troops from France and its allies. The resources of African countries are sought by various countries, each offering advantages to the local administration or trying to destabilize the supporters of the other.

Kazakhstan is particularly important for its oil, gas, uranium, and other resources. Possible political instability can greatly disrupt the international economy. Thus, a quick and harsh reaction from Russian troops is generated by Russia's interest in Kazakhstani reserves. Those resources are not pursued solely by Russia, so internal political conflicts must be viewed within an international context.

Regarding critical infrastructures, it must be understood that they refer to economic flows consisting of money, energy, information, products, and work. Thus, the globalization of the financial system represents a way to control money flows beyond the power of each state. Transport networks for oil, gas, electricity, uranium ore, and coal are being developed at the level of the entire planet. The information is supported by networks that include submarine cables to land networks and satellites. Moreover, individual terminals associated with social networks and with increased computing power, allow the accumulation of information at the individual level that have effects not yet fully known for the economy and the values of life. In addition, energy systems associated with information systems represent the biggest "machine" created by human society. The transport networks produced are developed at the planetary level and a disruption, for example, in the Suez Canal can affect the entire planet. Work is also an important flow, and the liberalization of work produces special effects such as migrations; see, e.g., the situation in the

EU, as well as elements of evolution and technological transfer that are sometimes unwanted, such as the example of China, which in the last 35 years has attracted the world's manufacturing based on cheap labor. At the same time, it has attracted technologies associated with manufactured products, leading to a major technological leap for this country.

It is important to note that critical networks are controlled. Money—through the reference interest rate of the central banks—works through the minimum wage that ensures the restoration of capacity for a new economic cycle, energy and information are not controlled at a centralized level (although partial control exists for fuel quantities and their prices in an indirect way), which leads to high volatility in these networks, and products are monitored through the World Trade Organization and various specific elements.

We will not talk about the intercorrelation between the various critical flows in order not to lengthen this analysis. However, the dynamics of the world economy are becoming nonlinear with effects that can be destructive in the short term and in areas that exceed the level of states or even economic and geographic regions.

This leads to an oscillating evolution of the economy that produces so-called crises periodically. The old biblical story of the seven fat cows and seven thin cows in the sense of the climatic influence on the food availability for the population becomes more complex due to the nonlinear effects that lead to crises due not only to climate changes but also to other causes related to the dynamics of the critical networks mentioned above. An important influence is the approach based on "fashion" principles. The development decision can be influenced by what are called "memes"—ideas that produce collective behavior not always based on optimality elements of human systems. An example is Germany's decision to close nuclear power plants, although this produces a short-term increase in the use of coal, which leads to adverse climate effects. Whoever makes decisions based on integrated considerations of evolution will have a competitive advantage in the medium and long term in sustainable development.

Influencing the providers of resources like gas and oil may be accomplished through various channels such as sociopsychological, religious, economic, and financial. As an example, a war in Israel-Gaza may trigger a drastic reduction in the supply of gas and oil from Arab countries to the rest of the world, with impacts on other ongoing conflicts, be they military or economical.

3.12 Conclusions

The organization of human economies into more efficient structures that create more GDP per unit of used energy has the effect of diminishing local system entropy. This generates an increase in the environmental temperature that is measurable if an irreversible thermodynamic model is used to describe the dynamics of energy intensity and the relation between these two interconnected, open systems.

The increase in temperature is evaluated based on the energy productivity (reverse of the energy intensity) data series for selected economies resulting in temperature increases of the same order of magnitude as the measured ones for the last 160 years. This leads to underlining the importance of the speed of temperature increases with regard to the recovery time constant of the environment.

The evolution of the temperature increase shows some dependency on periods of increased or decreased organization of the economies that should be more extensively analyzed. Additionally, it is suggested to prepare for the effects of extended conflicts expected to occur from climate change.

One important conclusion is that to compensate for the temperature increase associated with the entropy decrease due to new organization of economies it is important to consider the transfer of resources and management of waste, in such a manner as to diminish the temperature increase effect. Some suggestions are provided on the matter, especially regarding the concept of "circular economy."

In the next chapter we estimate climate change effect risks based on an analysis of big data series for temperature and precipitation within a geographic distribution.

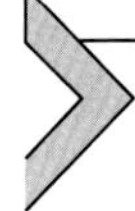

Annex 3.1. Food versus biofuels—an energy balance approach

Decision support analysis for allocating agricultural area between food and biofuels based on energy conservation

In recent years the production of fuels for transportation first saw an almost exponential increase then a sort of saturation from the adverse effects on agriculture for food products. Since both biofuels and agriproducts are expressed in energy units, we try here to find a balance in terms of energy units that are correlated to the land surface area available for food and biofuels in given economies, e.g., the USA, the EU, and China, to identify the optimal division of agricultural land to cover both basic food needs and those of biofuels without increasing food prices. In this first approach, food is measured in human daily energy intake converted to equivalent corn, and biofuel is measured in the fuel needed for cars, also converted into equivalent corn. These two corn quantities are then expressed in the land surface area needed to cultivate them, followed by a dynamic analysis of the optimal partition between feeding the population and supplying the cars.

Introduction

This analysis was done in 2013, and its presentation here is primarily as an example of applying an algorithm based on energy balance to decisions about using agricultural land resources.

According to The International Energy Outlook (2013), which presents an assessment by the US Energy Information Administration of the outlook for international energy markets through 2040, although liquid fuels—mostly petroleum-based—will remain the largest source of energy, the liquid share of world energy consumption will fall from 34% in 2010 to 28% in 2040 as projected high world oil prices lead many energy users to switch away from liquid fuels when feasible. The fastest-growing sources of world energy in the IEO 2013 assessment are renewables and nuclear power. The renewables share of total energy use will rise from 11% in 2010 to 15% in 2040, and the nuclear share will rise from 5% to 7%.

Rapid growth has been witnessed in recent years in the production and consumption of biofuels for powering combustion engines for the transportation economic sector. Remarkably, this trend is forecast to continue, and a further doubling is expected to occur in the present decade. The most important biofuels today are ethanol, based on cereals (e.g., corn) and sugar crops (e.g., sugarcane or sugar beet), and biodiesel, based on vegetable oils such as rapeseed, palm, soybeans, and canola oil. While ethanol and biodiesel have expanded into the existing markets and infrastructures of gasoline and diesel, other renewable fuels have begun to emerge as potentially viable alternatives, in particular biobutanol and biohydrogen. Furthermore, great expectations rest on cellulosic biofuels using wood, grasses, or organic waste (Jurgen Scheffran, Biomass for Biofuels: Strategies for Global Industries, Edited by Alain Vertes, Nasib Qureshi, Hans Blaschek and Hideaki Yukawa, John Wiley & Sons, Ltd., 2010, p. 27).

Biofuels are not a silver bullet for the energy problems of the world. To solve the issue of dwindling fossil fuel reserves, all viable means of harvesting energy should be pursued to the fullest, with biofuels remaining a reliable alternative energy resource. With more development and research, it is possible to overcome the disadvantages of biofuels and make them suitable for widespread consumer use. When the technology becomes available, many of the disadvantages will be minimized, and the market clearly has potential. Much of this could rely on the ability of energy producers to discover better agricultural plants to raise for fuel that use less water, require less land, and grow quickly.

No fuel source is completely positive or negative. Consumers must weigh the pros and cons of biofuels to determine whether they feel comfortable with this resource as an alternative to traditional fuels.

Advantages

Biofuel advocates frequently point out the advantages of these plant- and animal-based fuels, such as:

- Cost: Biofuels have the potential to be significantly less expensive than gasoline and other fossil fuels. This is particularly true as worldwide demand for oil increases, oil supplies dwindle, and more biofuel sources become apparent.
- Source material: Whereas oil is a limited resource that comes from specific materials, biofuels can be manufactured from a wide range of materials including crop waste, manure, and other by-products. This makes it an efficient step in recycling.
- Renewability: It takes a very long time for fossil fuels to be produced, but biofuels are much more easily renewable as new crops are grown and waste material is collected.
- Security: Biofuels can be produced locally, which decreases the nation's dependence on foreign energy. By reducing dependence on foreign fuel sources, countries can protect the integrity of their energy resources and keep them safe from outside influences.
- Economic stimulus: Because biofuels are produced locally, biofuel manufacturing plants can employ hundreds or thousands of workers, creating new jobs in rural areas. Biofuel production also increases the demand for suitable biofuel crops, providing economic stimulus to the agriculture industry.

- Lower carbon emissions: Biofuels produce significantly less carbon output and fewer toxins than fossil fuels when burned, making them a safer alternative that preserves atmospheric quality and lowers air pollution.

Disadvantages

Despite the many positive characteristics of biofuels, these energy sources also have many disadvantages:

- Food shortages: There is a concern that using valuable cropland to grow fuel crops could have an impact on the cost of food and lead to food shortages.
- Food prices: As demand grows for food crops such as corn for biofuel production, prices for necessary staple food crops could also rise.
- Energy output: Biofuels have lower energy output than traditional fuels and therefore require higher consumption to produce the same energy. This has led some energy analysts to believe that biofuels are not worth the work.
- Production of carbon emissions: Several studies have been conducted to analyze the carbon footprint of biofuels, and while they may be cleaner to burn, there are strong indications that the process to produce the fuel—including the machinery necessary to cultivate the crops and the plants to produce the fuel—produces hefty carbon emissions.
- High cost: Refining biofuels into more efficient energy outputs and building the necessary manufacturing plants to increase biofuel quantities often require high initial capital investments.
- Water use: Massive quantities of water are required to properly irrigate biofuel crops and manufacture the fuel, which could strain local and regional water resources.

Biofuel (bioethanol) markets in selected nations

The world's top ethanol fuel producers in 2013 were the United States with 13.3 billion gallons (BG), Brazil with 6.3 BG, Europe with 1.37 BG, and China with 696 million gallons.

In 2011, the world's top ethanol fuel producers, the United States with 13.9 BG (52.6 billion liters) and Brazil with 5.6 BG (21.1 billion liters), together accounted for 87.1% of the world production of 22.36 billion US gallons (84.6 billion liters) (Renewable Fuels Association (March 6, 2012) ("Accelerating Industry Innovation –2012 Ethanol Industry

Outlook, Retrieved March 18, 2012." pp. 3, 8, 10, 22, 23). Strong incentives, coupled with other industry development initiatives, are giving rise to fledgling ethanol industries in countries such as Germany, Spain, France, Sweden, China, Thailand, Canada, Colombia, India, Australia, and some Central American countries (http://en.wikipedia.org/wiki/Ethanol_fuel#Energy_balance, Retrieved 28 March 2014).

The United States, by using corn primarily as feedstock, produced 13.3 BG of ethanol in 2013 compared with 13.1 BG in 2012. Brazil's ethanol production using sugar cane totaled 6.3 BG in 2013, an increase of almost 11.2% over 2012. All of Europe produced 1.37 BG in 2013 compared with 1.14 BG in 2012. The EU produced 1.18 BG in 2010 compared with 1.04 BG in 2009, but from 2006 to 2007, ethanol production decreased by 2.2%, or 570 million gallons, because of higher feedstock prices. China produced 696 million gallons in 2013 compared with 555 million gallons in 2012. Chinese fuel ethanol production, using primarily corn, was 430 million gallons in 2007, an increase of less than 1% over 2006 (in 2007, total Chinese fuel ethanol production was 486 million gallons).

United States

The United States is the largest consumer of transportation energy in the world. It produces and consumes more ethanol fuel than any other country.

The development of efficient renewable energy sources has been a US policy goal since the first oil embargo of the early 1970s. While the more than two decades of progress have been slow, biofuel production and use have grown rapidly since the mid-1990s, driven by federal policies aimed at reducing air and water pollution (INFORMA, 2006). Moreover, military engagement and high energy costs provided new arguments for investing in biofuel technology as a means to (1) reduce dependence on oil imports, (2) lessen GHG emissions, and (3) create benefits for the agricultural economic sector.

The impetus to drive the implementation of biofuels is expressed at the highest levels of US economic and technology policies, as exemplified in the 2007 State of the Union address when the US president announced a planned increase in renewable fuels to 133 billion liters (35 BG) by 2017—that is, nearly five times the 2007 level Jurgen Scheffran, Biomass for Biofuels: Strategies for Global Industries, Edited by Alain Vertes, Nasib Qureshi,

Hans Blaschek and Hideaki Yukawa, John Wiley & Sons, Ltd., 2010, pp. 37–38).

The production of fuel ethanol from corn in the United States is controversial for a few reasons. Production of ethanol from corn is only 15% to 20% as efficient as producing it from sugarcane. Ethanol production from corn depends greatly on subsidies and consumes a food crop to produce fuel (J. K. Bourne JR, R. Clark, *Green Dreams*, National Geographic Magazine, October 2007 p. 41).

Based on current gasoline consumption, it would take 11 acres (45,000 m^2) of corn per person to replace the gasoline supply (Ziegler, Alexis (2006), Alternative Energy Sources. Greenhaven Press. p. 73). The subsidies paid to fuel blenders and ethanol refineries have often been cited as driving up the price of corn. At the same time, farmers have planted more corn and converted considerable land to corn (maize) production, which generally consumes more fertilizers and pesticides than other land uses (J.K. Bourne JR, R. Clark, *Green Dreams*, National Geographic Magazine, October 2007 p. 41). This is at odds with the subsidies paid directly to farmers, which are designed to take corn land out of production and pay farmers to plant grass and idle the land, often in conjunction with soil conservation programs, in an attempt to boost corn prices. Recent developments in cellulosic ethanol production and commercialization may allay some of these concerns. A theoretically much more efficient way of producing ethanol has been suggested that uses sugar beets, which make about the same amount of ethanol as corn without using the corn food crop, especially since sugar beets can grow in less tropical conditions than sugar cane (Kinver, Mark, September 18, 2006, Biofuels look to the next generation, BBC News. Retrieved August 27, 2011).

In October 2008, the first "biofuels corridor" was officially opened along I-65, a major interstate highway in the central United States. Stretching from northern Indiana to southern Alabama, this corridor consists of more than 200 individual fueling stations and makes it possible to drive a flex-fueled vehicle from Lake Michigan to the Gulf of Mexico with less than a quarter tank of fuel from an E85 pump (http://en.wikipedia.org/wiki/Ethanol_fuel_by_country

At the end of 2013, E15 was being offered at approximately 60 retail stations in 12 states—a nearly 10-fold increase since the beginning of 2013.

Additionally, as 2014 commenced, production was set to begin at the first wave of commercial cellulosic ethanol plants.

Amid the slow economic recovery from the Great Recession, the ethanol industry continued to have a profoundly positive impact on the fiscal health of rural America.

In 2013, the production of 13.3 BG of ethanol supported 86,504 direct jobs in renewable fuel production and agriculture, as well as 300,277 indirect and induced jobs across all sectors of the economy. Moreover, America's ethanol industry added $44 billion to the nation's gross domestic product and generated $8.3 billion in taxes. The sector's economic activity and job creation helped raise household income by $30.7 billion. Meanwhile, the US ethanol industry spent $36.1 billion on raw materials, other inputs, and goods and services. In just a few short decades, the ethanol industry's value of output has grown to surpass that of the internet publishing and broadcasting sector, farm machinery and equipment manufacturing, the snack food industry, and other major American industrial sectors.

US dependence on imported crude oil and petroleum products is plunging to depths not seen since the early 1990s. After peaking at 60% in 2005, import dependence has fallen steadily and registered at an estimated 35% in 2013.

The surge in ethanol production has reduced gasoline imports from 600,000 barrels per day in 2005 to near zero today. Looked at another way, the ethanol produced in 2013 displaced the amount of gasoline refined from 462 million barrels of imported crude oil. That is roughly equivalent to the amount of crude oil imported annually from Venezuela and Iraq combined.

While tremendous strides have been made to reduce dependence on imports of refined petroleum products like gasoline, progress has been slower in reducing raw crude oil imports. Imports accounted for an estimated 51% of the crude oil processed by US refineries in 2013. The Department of Energy's long-term projections suggest imports will continue to make up more than half of US crude oil supplies at a cost to the American economy of roughly $1 billion per day. And because crude oil is a global commodity, the recent increase in US fracking has not resulted in lower oil prices. Against this backdrop, continued growth in the production of ethanol will remain vitally important as a strategy for diversifying the fuel market and improving domestic energy security.

European Union

Like the United States, the EU aims to reduce its dependence on external energy sources and create a new stimulus for the rural economy (Faaij,

2006). High oil prices and the ratification of the Kyoto Protocol in 2005 have provided additional incentives to strongly promote alternative fuels.

The predominant feedstocks used for ethanol production in the EU are sugar beets and wheat. In addition, surplus wine has been converted into ethanol, while corn and potatoes have also been used. A major limiting factor in converting the large diversity of feedstocks into ethanol is the costs of the feedstocks themselves. However, these costs may be reduced in the short to mid term by implementing cellulosic ethanol or more simply by implementing economically improved agricultural practices (Yürgen Scheffran, Biomass for Biofuels, The Global Demand for Biofuels: Technologies, Markets and Policies, John Wiley & Sons, Ltd., 2010, pp. 40, 41).

Because of the constraints of ethanol for fuel production, given the currently available technologies and feedstocks, the EU is also placing a large effort on developing second-generation (i.e., cellulosic) biofuels. Many countries in the EU are also investigating alternative routes based on biomass gasification for syngas production and conversion to biofuels; that is, methanol, dimethylester, Fischer—Tropsch liquids, or hydrogen (Faaij, 2006).

Due to concerns over the global food crisis and biofuel sustainability, the EU is considering a ban on certain types of biofuels, especially those grown on vulnerable lands (NYT, 2008). While some governments have begun rolling back biofuel subsidies, others favor high import tariffs so that the European biofuel industry could implement a dynamic biofuel value chain starting from the production of biomass feedstocks in the EU (Yurgen Scheffran, Biomass for Biofuels, The Global Demand for Biofuels: Technologies, Markets and Policies, John Wiley & Sons, Ltd., 2010).

The flagship initiative for a resource-efficient Europe under the Europe 2020 strategy supports the shift toward a resource-efficient, low-carbon economy to achieve sustainable growth. Increasing the share of energy from renewable sources is part of a decisive move toward a low-carbon economy.

In the EU-27 the share of energy from renewable sources in gross final energy consumption increased from 8.1% in 2004 to 12.5% in 2010.

The year 2010 marks the end of the first third of the period running from the base year 2005 until 2020, for which Directive 2009/28/EC has set a 20% target at the EU level. At the same time, we can observe an increase from the base year value of 8.5%—12.5% in 2010—which is roughly one-third of the gap between the base year share of renewables and the 20% target. Thus, at the aggregated EU level, progress toward the 2020 target is in line with the expected linear trajectory of progress toward the target.

For all EU countries, there is a common 2020 target of 10% for the share of renewable energy in the transport sector. Directive 2009/28/EC stipulates that only biofuels/bioliquids that fulfill sustainability criteria should be included.

Total fuel consumption of road transport (as well as total transport) fell from its peak in 2007 and continued its downward trend to 2010. In 2010, based on the accounting principles described in the Directive, the share of energy from renewable sources in all modes of transport reached 4.7% in the EU-27.

In 2010 total electricity use in transport was 3.4% below its use in 2004. Consumption by electric road vehicles continued to increase but was still negligible. Nonroad electricity consumption decreased slightly more than total transport electricity use (−3.5% during the period). The use of transport biofuels reported as sustainable (compliant with the rules in the Directive) increased considerably over the 2004−2010 period.

Fig. 3.1 shows the unadjusted consumption of energy sources in transport. The percentages in Table 3.1 are adjusted for the rules in the Directive (Marek Šturc, Environment and Energy, Eurostat−Statistics in Focus 44/2012, pp. 1, 6).

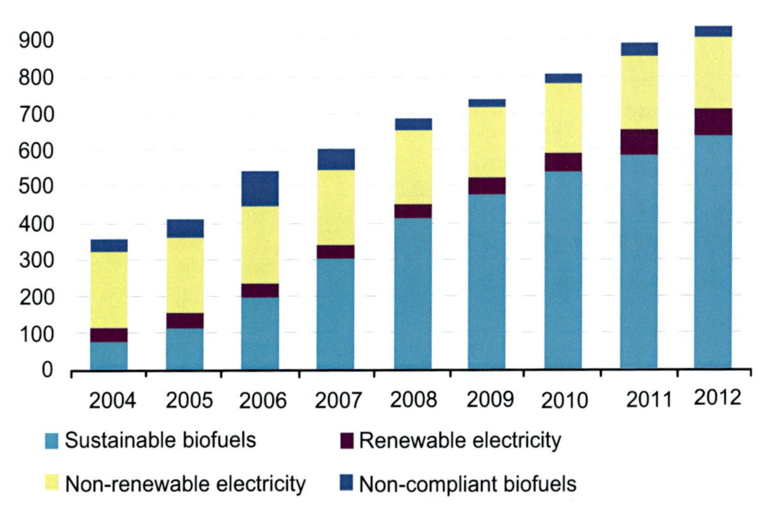

Figure 3.1 Consumption of electricity and biofuels in transport (in petajoules). *Source: Marek Šturc, Environment and Energy, Eurostat−Statistics in Focus 44/2012; Eurostat (SHARES 2010 application).*

Table 3.1 Share of renewable energy sources in transport (in percent).

		2007 (%)	2008 (%)	2009 (%)	2010 (%)	2011 (%)
EU27	**EU27**	**2.8**	**3.5**	**4.3**	**4.8**	**3.4**
BE	**Belgium**	1.3	1.3	3.3	4.1	4.0
BG	**Bulgaria**	0.4	0.5	0.5	1.0	0.4
CZ	**Czech Republic**	1.0	2.3	3.7	4.6	0.7
DK	**Denmark**	0.3	0.3	0.4	0.9	3.8
DE	**Germany**	7.4	6.0	5.5	6.0	5.9
EE	**Estonia**	0.1	0.1	0.2	0.2	0.2
IE	**Ireland**	0.5	1.3	1.9	2.4	3.9
EL	**Greece**	1.2	1.0	1.1	1.9	0.7
ES	**Spain**	1.2	1.9	3.5	4.7	0.4
FR	**France**	3.6	5.7	6.2	6.2	0.5
IT	**Italy**	0.8	2.3	3.7	4.6	4.7
CY	**Cyprus**	0.1	1.9	2.0	2.0	0.0
LV	**Latvia**	0.9	0.9	1.2	3.3	3.5
LT	**Lithuania**	3.7	4.2	4.3	3.6	3.7
LU	**Luxembourg**	2.1	2.1	2.1	2.0	2.1
HU	**Hungary**	1.0	4.0	4.2	4.7	5.0
MT	**Malta**	0.0	0.0	0.0	0.0	0.0
NL	**Netherlands**	2.9	2.7	4.3	3.1	4.6
AT	**Austria**	6.3	7.5	8.9	8.6	7.6
PL	**Poland**	1.2	3.6	5.1	6.3	6.5
PT	**Portugal**	2.4	2.4	3.9	5.6	0.4
RO	**Romania**	1.8	2.6	3.5	3.1	2.0
SI	**Slovenia**	1.1	1.5	1.9	2.8	2.1
SK	**Slovakia**	3.5	3.9	4.9	4.8	5.0
FI	**Finland**	0.4	2.4	4.0	3.8	0.4
SE	**Sweden**	5.7	6.3	6.9	7.2	9.4
UK	**United Kingdom**	1.0	2.1	2.7	3.1	2.7

Source: Marek Šturc, Environment and Energy, Eurostat—Statistics in Focus 44/2012; Eurostat (SHARES 2010 application).

China

Facing rapidly growing demand for transportation fuels (with an expected consumption of 228 MT in 2020), China set a target of producing 11 MT of biofuels from renewable sources by 2020 (INFORMA, 2006). At the Renewables 2004 conference in Germany, China announced a national commitment to obtain 16% of the country's energy from renewables by 2020 (Martinot and Junfeng, 2008). Notably, fuel ethanol is exempt in this country from consumption tax and value-added tax; moreover, several provinces have introduced compulsory ethanol-blended gasoline. However, despite a large energy demand, food crops are a priority for land use in China. Because of serious sugar shortages, ethanol production from sugarcane was stalled in August 2005 (INFORMA, 2006). The biodiesel program is less developed than the bioethanol program, and only a few small plants are operating and mainly use waste cooking oil or oilseeds as feedstock (INFORMA, 2006).

China is promoting ethanol-based fuel on a pilot basis in five cities in its central and northeastern region, a move designed to create a new market for its surplus grain and reduce the consumption of petroleum. The cities include Zhengzhou, Luoyang, and Nanyang in central China's Henan province and Harbin and Zhaodong in the Heilongjiang province of northeast China. Under the program, Henan would promote ethanol-based fuel across the province by the end of 2023. Officials say the move is of great importance in helping to stabilize grain prices, raise farmers' incomes, and reduce petrol-induced air pollution (http://en.wikipedia.org/wiki/Ethanol_fuel_by_country).

Corn for food or for cars?

Most studies on energy use in the transportation sector emphasize the growing importance of automobiles in individual transportation and the continued dependence on fossil fuels.

In rich industrialized countries, biomass represents on average about 3% of the total amount of primary energy carriers. In emerging markets, it accounts for 38%. In some particularly poor countries, it reaches more than 90% (Dieter Deublein & Angelika Steinhauser—Biogas from Waste and Renewable Resources, Wiley-VCH Verlang GmbH & Co. KGaA, Weinheim, Germany, 2008).

In 2007, global biofuel production amounted to 62 billion liters (GL: gigaliters) or 16.4 BG per year, corresponding to 36 million tons of oil equivalent or 1.8% of total global transport fuel consumption in energy terms (OECD, 2008).

Fuel ethanol accounts for most of the world's biofuels, with a production of 49.6 GL (13.1 BG) in 2007, a dramatic rise compared with about 20 GL in 2002. Almost half of the ethanol is produced in the United States, 38% in Brazil, 4.3% in the European Union, and 3.7% in China. While Brazil was by far the world's largest producer throughout the 1980 and 1990s, in about 2005 Brazil was overtaken by the US, although it remains the largest exporting country.

For every 10 ears of corn grown in the United States today, only two are consumed directly by humans as food. The remaining eight are used in almost equal shares for animal feed and ethanol. In the 12 months from August 2011 to August 2012, the US biofuels industry used more corn for fuel than domestic farmers did for livestock feed—a first for the industry. This significant milestone in the shifting balance between crops for food versus fuel shows the impact of government subsidies for the biofuels industry. It could also represent a tipping point in the conflict between food and fuel demand in the future.

The use of ethanol for fuel has had a damaging impact on food markets, especially in poorer countries. In the United States, ethanol is mostly made from yellow corn, and as the market for alternative fuels boomed, the price of yellow corn went up. Many farmers saw the potential to make more money and switched from white to yellow corn. White corn is the main ingredient of tortillas in Mexico, and as the supply of white corn dropped, the price doubled, making the base of most Mexican foods unaffordable. Many people see this as unacceptable and want no overlap between food crops and fuel crops. Others point out that the earth is thought to be able to support double the current human population and press that available resources, such as unused farmable land, should be better handled.

The Renewable Fuels Association confirms that ethanol production does in fact increase the price of corn through increased demand. It cites this as a positive economic effect for US farmers and taxpayers but does not elaborate on the effect for populations where field corn is a part of the staple diet.

On March 9, 2011, Senator Dianne Feinstein from California introduced a bill that repealed the corn subsidies in the United States. She is quoted as telling Congress, "Ethanol is the only industry that benefits from a triple crown of government intervention: its use is mandated by law, it is protected by tariffs, and companies are paid by the federal government to use it. It's time we end this practice once and for all."

http://en.wikipedia.org/wiki/Corn_ethanol.

Over the past year, US farmers used 5 billion bushels of corn for animal feed and residual demand. During that time, the nation used more than 5.05 billion bushels of corn to fill its gas tanks. And while some of the corn used to produce these biofuels will be returned to the food supply (as animal feed and corn oil), a large proportion will be solely dedicated to our gas tanks.

According to Rabobank's head of agricultural research, Luke Chandler, this shift in the balance between food and fuel could be the tipping point in world grain markets. China, once able to supply its internal corn demand, currently expects to import (from the USA) a few million tons of corn per year. This will likely place additional stress on the US corn industry, as it will introduce another source of demand (and corresponding market pressures) for the nation's corn harvests.

The basic approach taken is energy balance decision support analysis for allocating agricultural area between food and biofuels based on energy conservation

Although all food has a tag with information on the amount of energy, usually in kcal, and biofuels are expressed in energy equivalent units, we know of no attempt yet to make a balance between the two to have a common measure related to finding an optimal partition of agricultural land to ensure enough food for the population and assess how many cars may be supplied with biofuel from the remaining land.

To do this, we start with the data in Annex 3.1.1, which gives the amount of energy associated with various food products. Obviously, one may discuss at length the associated energy content, but our purpose is to show how the problem may be approached. Therefore, we do not go into these details in this paper but simply present a synthesis of various methods in Table 3.2.

As we see the Atwater conversion factor is giving an energy intake per adult-day of 2739 kcal. We consider this the basic data for our calculation of food energy needs for the population of the USA, the EU, and China. China's population was the largest of any country in the world at the time of this analysis, the EU was the third-most populous geographic region after China and India, and the USA was the fourth-most populous region in the world (and the third-most populous country).

To continue with the calculations, we take the value of corn energy (total carbohydrates), i.e., 4.03 kcal/g, from Annex 3.1.1. This value shows that the intake per adult-day of energy may be expressed in corn equivalent as $2739/4.03 = 679.65$ g. For a year (365 days), this amounts to 248,073.2 g or 248.07 kg/adult.

For the entire population of the selected nations (approximately 317 M people for the USA, 505 M for the EU, and 1363 M for China) (http://en.wikipedia.org/wiki/List_of_countries_by_population, http://ec.europa.eu/eurostat), corn equivalent production is 78.6, 125.3, and 338.1 Mt, respectively.

Considering a production of 4000 kg corn/hectare, the estimated agricultural land surface needed to feed the total population is 19.7 Mha for the USA, 31.3 for the EU, and 84.5 Mha for China.

Table 3.2 Methods for determining energy content of foods.

Energy conversion factor	Description			Per adult-day energy consumption		Difference in prevalence of low energy intake
	Protein based on	Carbohydrates by difference	Energy from fiber	*Kcal*	%	
Atwater	Jones	Total	Included	*2739*	101.2	−1.8
ME2	Jones	Available	Included	*2714*	100.3	−0.6
Merrill and watt	Jones	Total	Included	*2706*	100	0
ME1	Jones	Available	Ignored	*2698*	99.7	0.2
$NME2^{AA}$	Total AA	Available	Included	*2634*	97.3	3.3
$NME2^{Jones}$	Jones	Available	Included	*2632*	97.3	3.4
$NME2^{6.25}$	6.25	Available	Included	*2631*	97.2	3.5
$NME1^{AA}$	Total AA	Available	Ignored	*2621*	96.9	4.1
$NME1^{Jones}$	Jones	Available	Ignored	*2619*	96.8	4
$NME1^{6.25}$	6.25	Available	Ignored	*2618*	96.7	4.1

Source: Wikipedia.

With this, we have arrived at an evaluation of the land surface needed to feed the population. Let us see how much surface is needed to "feed" the existing cars in these countries.

For the United States, a recent study (John W. Bickham[1] and Mark A. Thomas[2], **Eco-Environmental Impact of Bioenergy Production,** Journal of Resources and Ecology 1(2):110–116. 2010), provides the number 3000 G.car.miles/year (240 M vehicles at the same km/year amount). Considering that each vehicle travels 20,000 km/year and consumes an average of 8L/100 km of gasoline, we have a consumption per car per year of 1600 L. Thus, annual consumption for the USA is 384 GL.

The number of vehicles in the EU is about 256 million. Considering the same rate of consumption, annual consumption in the EU is 410 GL. The number of vehicles in China is about 78 million (http://en.wikipedia.org/wiki/Motor_vehicle#European_Union). At the same rate of consumption, annual consumption in China is 125 GL.

The provides a synthesis of the above data.

Biofuel and agriculture-related data

Data\country	*EU*	*USA*	*China*
Food/year/population (ha)	31319241.63	19659801.18	84530943.24
Population #	505000000	317000000	1363000000
Vehicles #	256000000	240000000	78000000
Surface for total vehicles (ha)	480768000	450720000	146484000
Country surface (km^2)	4296752	9161966	9598089
Arable land (%)	40.11	16.29	14.65
Arrable land (km^2)	1723427.227	1492484.261	1406120.039
Arrable land (ha)	172342722.7	149248426.1	140612003.9
Surface available for vehicles (ha)	141023481.1	129588625	56081060.61
Surface for vehicles at 2× surface for population food	109704239.5	109928823.8	−28449882.63
Biofuel in car fuel (%)	22.82%	24.39%	−19.42%

To convert into corn equivalent, we assume that consumption equates to 1.5 L ethanol/L gasoline (ethanol has lower calorific power than gasoline) and that the production of 1 L of ethanol requires 3.13 kg of corn. Thus the required quantity of corn is 1802.88 Gkg for the USA, 1924.95 Gkg for the EU, and 586.88 Gkg for China. Considering the same production of 4000 kg/ha, the surface needed to supply fuel for all the vehicles would be 450.72 Mha in the USA, 481.24 Mha in the EU, and 146.72 Mha in China.

The arable land surface in the USA is 16.29% of the total surface area of 9,161,966 km^2, i.e., 149.25 Mha. The surface area required for food and biofuels in the USA is 19.7 + 450.72 = 470.42 Mha. The arable land surface in the EU-27 is 40.11% of the total surface area of 4296752 km^2, i.e.,172.34 Mha (Eurostat—online data codes: demo_r_d3area, ef_oluft and ef_lu_ovcropaa). The surface area required for food and biofuels in the EU-27 is 31.3 + 481.24 = 512.54 Mha. The arable land surface in China is 14.65% of the country's total surface area of 9,598,089 km^2, i.e., 140.63 Mha (http://en.worldstat.info/Asia/China/Land). The surface area required for food and biofuels in China is 84.5 + 146.72 = 231.22 Mha.

In the USA, the total surface area required for food and transportation of 470.42 Mha is much larger than the arable land area of 149.25 Mha. The same applies to the EU, where the total surface area required for food and transport is 512.54 Mha versus a total surface area of 172.34 Mha. For China, the total surface area required for food and transport is 231.22 Mha versus a total surface area of 140.63 Mha.

It is clear, though, that an economy cannot rely on meeting only basic food needs, so it is normal to consider that a reserve quantity should be set aside every year to cover contingencies related to food availability. In addition, there is the need to feed livestock, as mentioned above. If we consider that this quantity should cover 2 years of food needs, then the required surface area for the USA will be 2 × 19.7 Mha = 39.4 Mha, the surface area for the EU will be 2 × 31.3 Mha = 62.6 Mha, and the surface area for China will be 2 × 84.5 = 169 Mha.

Optimal partition of arable land between food and biofuel crops

Considering the calculations above we see that the surface of land needed for food is smaller than the total arable surface of the USA and the EU. In contrast, for China the surface area of land needed for food is larger than the total arable surface. This means finding the optimal partition of the surface area between that needed for food and for biofuel. The surface area that remains for biofuels after meeting required food needs is smaller than the one required to provide biofuel for all the vehicles in each country, and the result of the optimal calculation provides either the number of cars that may be fully supplied with biofuels or the percentage of biofuel in all vehicle fuels.

The basic logic for the optimization is that the less surface is available for population food (doubled for risk coverage purposes), then, more is available

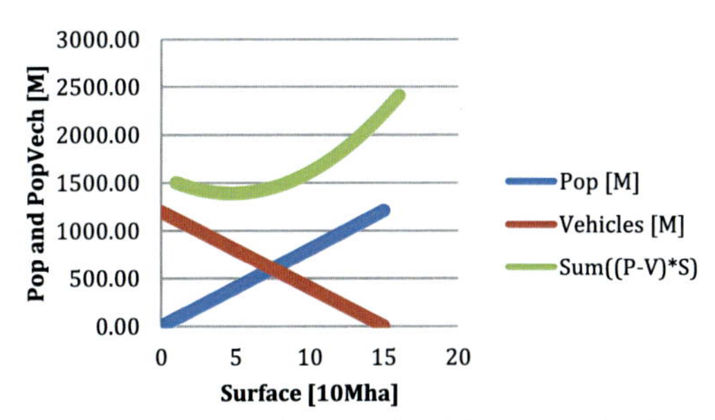

Figure 3.2 USA, optimal partition of arable land between food and biofuel. *Source: Author's calculations.*

for biofuel, and there is a minimum value of the difference of the surfaces under the two lines that gives the optimal surface partition. The vertical axis measures the population equivalent (i.e., double the population for food and multiply the number of vehicles by 15 for biofuel).

The calculation for the USA, as shown in Fig. 3.2, provides an optimal surface area of about 40 Mha for food and 59 M vehicles, or about 24% of the total number; thus, the available land surface area for biofuels may supply about 24% of the total fuel consumption of the vehicles in the USA.

The graphs for the EU and China are straightforward and are not presented here.

Conclusions

The approach we have taken to assess growing corn on available agricultural land, partitioned between feeding the existing population of a country and producing biofuel for cars, leads to several primary conclusions.

The first is that feeding the population is paramount to anything else, and ensuring the necessary quantities, including risk coverage, determines the basic agricultural land surface to keep untouched by alternative uses.

The second is the possibility of determining how much biofuel may be available from the remaining agricultural land, such as determining the size of the car fleet that could be supplied for a specific economy with that amount of biofuel.

The third conclusion provides a way to evaluate the percentage of biofuel in the total fuel supply. This is determined based on the existing number of cars in the country and the available biofuel versus the need for fuel for those cars. Obviously, if the supply of biofuel is greater than the need, the exporting option is opened, which is not the case for either EU or the United States.

The fourth conclusion is related to the export of food that exceeds the need (including basic reserves).

For the United States, the percentage of biofuel is 24%, which signifies the greater motorization of that economy. The EU is at the same level of 23%, while China must import to meet about 19% of its biofuel needs if it wants to feed its huge population.

Finally, we note that our approach is basic and must be extended to consider the rotation of basic cultures (corn, wheat, and sunflower), crops for animal stock, and the technological changes needed to fully use biofuels in car motors. Moreover, a thorough evaluation of GHG emissions should be performed given that agriculture is both a source and a sink of emissions.

Annex 3.1.1

	Protein kcal/g (kJ/g)	Fat kcal/g (kJ/g)	Total carbohydrate kcal/g (kJ/g)[§]
Cornmeal, whole ground	*2.73* (11.4)	*8.37* (35.0)	*4.03* (16.9)

References

Dieter Deublein & Angelika Steinhauser—Biogas from Waste and Renewable Resources, 2008. Wiley-VCH Verlang GmbH & Co. KGaA, Weinheim, Germany.

Du, X., Graedel, T., 2011. Global in-use stocks of the rare Earth elements: a first estimate. Environmental Science and Technology 45 (9), 4096–4101.

EASAC, 2016a. Indicators for a Circular Economy.

EASAC, 2016b. Priorities for Critical Materials for a Circular Economy.

Faaij, A., 2006. Bio-energy in Europe: changing technology choices. Energy Policy 34, 322–342.

Hagelüken, C., 2014. Recycling of critical metals. In: Critical Metals Handbook. John Wiley.

INFORMA, March, 2006. The Emerging Biobased Economy: A Multi-Client Study Assessing the Opportunities and Potential of the Emerging Biobased Economy, Developed by Informa Economics, Inc. In Participation with MBI International and the Windmill Group.

Martinot, E., Junfeng, L., 2008. Powering China's development: the role of renewable energy, excerpts from the Worldwatch special report: powering China's Development: the Role of Renewable Energy (Washington, DC, November 2007, 50 pp.). Renewable Energy World Magazine 11 (1). January/February 2008; full report available at. http://www.worldwatch.org/node/5491.

NYT, 2008. Europe May Ban Imports of Some Biofuel Crops. New York Times, 15 January.

OECD, 2008.

Scheffran, Y., 2010. Biomass for Biofuels—The Global Demand for Biofuels: Technologies, Markets and Policies. John Wiley & Sons, Ltd.

M. Šturc, Environment and Energy, Eurostat–Statistics in Focus 44/2012.

Sverdrup, H., Ragnarsdottir, K., 2014. Natural resources in a planetary perspective. Geochemical Perspectives 3, 129–336.
USEA/USAID (Ed.), June 1999. USEA/USAID Handbook of Climate Change Mitigation Options for Developing Country Utilities and Regulatory Agencies. USEA/USAID, USA.
Wellmer, F., Hagelüken, C., 2015. The feedback control cycle of mineral supply, increase of raw material efficiency, and sustainable development. Minerals 5, 815–836.

Further reading

Angrick, M., Burger, A., Lehmann, H. (Eds.), 2014. Factor X: Policy, Strategies and Instruments for a Sustainable Resource Use. Springer, Berlin.
Ayres, R., Peiró, L., 2013. Material efficiency: rare and critical metals. Phil Trans R Soc A 371: 20110563 CWIT (2015). In: Countering WEEE Illegal Trade. http://www.cwitproject.eu/wp-content/uploads/2015/09/CWIT-Final-Report.pdf.
Bakchi, B.,R., et al. (Eds.), 2012. Thermodynamics and the Destruction of Resources. Cambridge University Press, Cambridge, ISBN 9780521884556.
Berger, S. (Ed.), 2009. The Foundations of Non-equilibrium Economics. Routledge Advances in Heterodox Economics, New York, ISBN 9780415777803.
Colombo, U., 2000. Energia. Storia e scenari. Universale Donzelli, Roma.
Corliss, W.,R., 1968. Direct Conversion of Energy, Understanding the Atom. USAEC.
Daxbock, I., et al., 2013. The Great Helium Escape. University of Iceland.
de Groot, S.R., Mazur, P., 1984. Non-Equilibrium Thermodynamics. Dover Books on Physics, ISBN 0486647412.
Deloitte, 2015. Study on Data for a Raw Material System Analysis: Roadmap and Test of the Fully Operational MSA for Raw Materials. Prepared for the European Commission, DG GROW.
EASAC, 2015. Circular Economy: Commentary from the Perspectives of Natural and Social Sciences.
EC, 2008. The Raw Materials Initiative—Meeting Our Critical Needs for Growth and Jobs in Europe. COM 699.
EC, 2010. Ad-Hoc Working Group on Defining Raw Materials.
EC, 2011. Tackling the Challenges in Commodity Markets and on Raw Materials. COM 25.
EC, 2013. Sustainable use of phosphorus. Consultative Communication.
EC, 2014. Report on critical materials for the EU. Report of the Ad hoc Working Group on defining critical raw materials.
EC, 2015. Closing the Loop—An EU Action Plan for the Circular Economy.
EC, 2015a. Circular Approaches to Phosphorus: From Research to Deployment.
EC, 2015b. M/543 Commission Implementing Decision C, vol 2015, p. 9096.
EIP, 2016. European innovation partnership on raw materials. Raw Materials Scoreboard.
Emily E. A., U.S. Carbon dioxide emissions down 11 percent, EcoWatch.
ETC/SCP, 2011. Green economy and recycling in Europe. http://scp.eionet.europa.eu/publications/2011_wp5.
http://www.fao.org/docrep/006/Y5022E/y5022e04.htm.
FAO, 2008. Biofuels: Prospects, Risks and Opportunities, the State of Food and Agriculture. UN Food and Agriculture Organization.
Fernandes, T., et al., 2015. Closing domestic nutrient cycles using microalgae. Environmental Science and Technology 49 (20), 12450–12456.
Fisher, B., 2008. Review and Analysis of the Peak Oil Debate. Institute for Defense Analyses. Report D-3542.
Graedel, M., et al., 2012. Methodology of metal criticality determination. Environmental Science and Technology 46, 1063–1070.

Guminski, K., 1964. Termodinamica Proceselor Ireversibile, Editura Academiei, Bucuresti.
Hagelüken, C., 2012. Recycling and substitution. Opportunities and limits to reduce the net use of critical materials in hi-tech applications. In: Proceedings of the Critical Materials Conference, Brussels, Belgium, March 2012.
Hagelüken, C., Meskers, C., 2010. Complex life cycles of precious and special metals. In: Thomas, E.G., Voet, E.van der (Eds.), Strüngmann Forum Report, Linkages of Sustainability.
ISF, 2010. Peak Minerals in Australia: A Review of Changing Impacts and Benefits. Institute for Sustainable Futures.
JRC, 2013. Critical metals in the path towards the decarbonization of the EU energy sector. Assessing Rare Metals as Supply-Chain Bottlenecks in Low-Carbon Energy Technologies.
JRC, 2016. Assessment of the Methodology on the List of Critical Raw Materials; Background Report.
Kerr, R., 2014. The coming copper peak. Science 343, 722–724.
Krautkraemer, J., 2005. The economics of natural resource scarcity, the state of the debate. Chapter 3 in scarcity and growth revisited. Resources for the Future.
Liu, Y., et al., 2008. Global phosphorus flows and environmental impacts from a consumption perspective. Journal of Industrial Ecology 12, 229–247.
Marscheider-Weidemann, F., et al., 2016. Rohstoffe für Zukunftstechnologien 2016.
Matsubae, K., Nagasaka, T., 2014. Resource logistics analysis on phosphorus and its application for science technology and innovation policy. In: Norton, T., Li (Eds.), Topical Themes in Energy and Resources. Springer, pp. 159–176.
Melissa C., L., Scientific American, October 7, 2011. The U.S. Now Uses More Corn for Fuel than for Feed.
Mihelcic, J., et al., 2011. Global potential of phosphorus recovery from human urine and faeces. Chemosphere 84, 832–839.
Moss, R., et al., 2011. Critical Metals in Strategic Energy Technologies—Assessing Rare Metals as Supply-Chain Bottlenecks in Low-Carbon Energy Technologies. http://setis.ec.europa.eu/newsroom-items-folder/announcement-new-report-on-critical-metals-in-strategic-energytechnologies.
Nurmi, P., et al., 2010. Finland's Mineral Strategy. http://www.mineraalistrategia.fi/etusivu/fi_FI/etusivu/_files/84608401427464240/default/FinlandsMineralsStrategy.pdf.
Nuttall, W., et al., 2012. Stop squandering helium. Nature 24, 573–575.
OECD, 2009. A Sustainable Materials Management Case Study—Critical Metals and Mobile Devices.
Parliament Office of Science and Technology Notes, 2007. Climate Change, UK.
Purica, I., 1992. Environmental change and the perception of Energy system dynamics. In: ICTP-trieste, Conference on "Global Change and Environmental Considerations for Energy System Development", Proceedings.
Purica, I., 2010. Nonlinear Models for Economic Decision Processes. Imperial College Press, London, ISBN 9781848164277.
REconserve, 2012. Rare Earths from Urban Mining. Tokyo.
Reuter, M., et al., 2015. Lead, Zinc and their minor elements: enablers of a circular economy. World of Metallurgy—Erzmetall 68 (3).
Schuler, D., et al., 2011. Study on Rare Earths and Their Recycling. http://www.oeko.de/oekodoc/1112/2011-003-en.pdf.
Steffen, W., et al., 2015. Planetary boundaries: guiding human development on a changing planet. Science 347. https://doi.org/10.1126/science.1259855.
Sverdrup, H., Koca, D., 2016. A Short Description of the WORLD 6.0 Model and An Outline of Elements of the Standard Parameterization. SIMRESS.

TNO, 2012. Second Trilateral EU-Japan-US Workshop.
Toninelli, P.A., October 2008. Energy Supply and Economic Development in italy: The Role of the State-Owned Companies, WORKING PAPER SERIES No. 146. Dipartimento di Economia Politica Università degli Studi di Milano—Bicocca. http://dipeco.economia.unimib.it.
Umicore, 2007. Electronic Scrap Recycling at Umicore.
UNEP, 2011. Assessing mineral resources in society: metal stocks and recycling rates. Summary Booklet Based on the Two Reports of the Global Metal Flows Group 'Metal Stocks in Society: Scientific Synthesis' and 'Recycling Rates of Metals: A Status Report'.
UNEP, 2013. Metal Recycling: Opportunities, Limits, Infrastructure.
Verhoef, E., et al., 2004. Process knowledge, system dynamics and metal ecology. Journal of Industrial Ecology 8, 23–43.
WEF/RRN, 2011. The political and economic implications of resource scarcity. http://www.weforum.org/content/pages/political-and-economicimplications-resource-scarcity.

CHAPTER FOUR

Big data analysis for climate change proof and risk mitigation

If a person sits half on ice and half on a hot plate, on the average he is O.K.
Grigore Moisil.

This fourth chapter presents a case analysis of climate change (CC) event risk maps based on big data series for temperature and precipitation with a geographical distribution for Romania. Risk maps are presented and applied to determine an impact on the Romanian gas network. Also presented is an application for the risk map of the Italian gas network subject to earthquakes and landslides. The possibility to introduce an insurance policy to CC risks is also determined based on the risk maps, as well as a better allocation of investments to mitigate or adapt to those risks.

CC is characterized by an increase in temperature and a larger standard deviation of the temperature distribution. With monthly temperature data starting in 1961 for each of the 40 regions of Romania, a twofold analysis was performed: first to check the evolution of the average temperature and the associated distribution's standard deviation and second to assess the risks stemming from the combined effects of temperature and precipitation in each region that result in flood, drought, snow, and freezing risks. These associated risks combine into a total risk, for which a potential insurance system is proposed that also correlates with the volatility of the temperature for a 10-year period within each region. The results of this analysis show evidence of a CC process and suggest the validity of a risk mitigation policy.

4.1 Introduction

Recent years have again raised the problem of CC as both a process that needs comprehensive agreement by world governments and a cause for a change in the risk patterns induced by changes in temperature and precipitation conditions.

To examine CC, several analyses were performed, especially looking at temperatures over past thousand to several hundred thousands of years

Climate Change and Circular Economics
ISBN: 978-0-443-29969-8
https://doi.org/10.1016/B978-0-443-29969-8.00005-7

(evaluated based on, e.g., ice specimens drilled from the Antarctic). Alternatively, temperature data were gathered for the last approximately 150 years from the beginning of the industrial era. The combined effect of temperature and precipitation had not been evaluated thus far to our knowledge, although the process of CC influences both parameters.

4.2 Data series

To analyze this type of process, a large amount of data is needed to capture long-term trends from the usual short-term noise. In the case of this paper, the data from Romania are provided by the Romanian National Meteorological Administration (NAM) and have both space and time discretization. The space values are given for each region of the country (40 regions), while the time data range from 1961 until 2011, with monthly average values for each region.

The time series distributions are assessed using @Risk for several 10-year periods in each case, with the averages and dispersions plotted in Annex 1. As an example, we will now analyze the data for one region.

The region is Prahova, situated north of Bucharest (the capital city). The temperature distributions are given in Fig. 4.1 for each 10-year period, starting in 1961. Note that the last period starts in 2001 and ends in 2013.

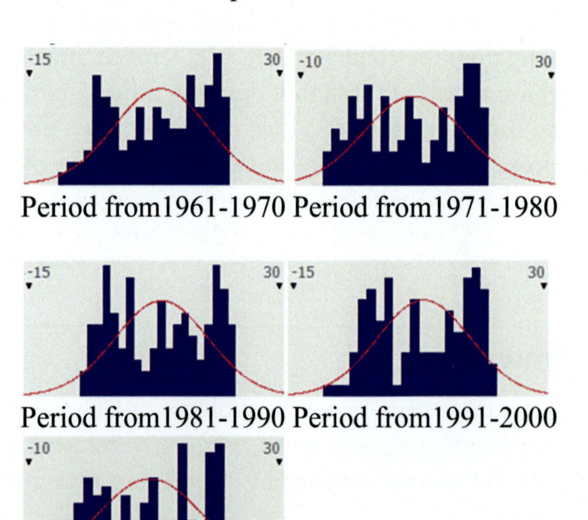

Period from 2001-2013

Figure 4.1 Prahova region temperature distributions.

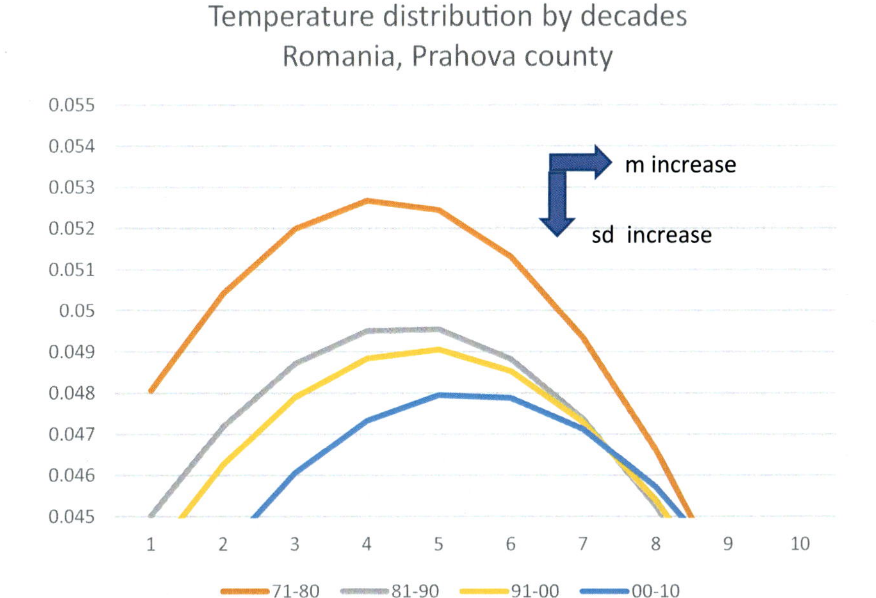

Figure 4.2 Temperature distribution comparisons.

The superposition of the distribution maxima, as presented in Fig. 4.2, shows without doubt the typical situation expected to characterize CC occurrence, i.e., increases in the average and the standard deviation. This translates into increased temperature values and occurrences with a higher frequency of extremes. Based on this result, the risks induced by the combined CC parameters of temperature and precipitation will be evaluated.

4.3 Risk mapping by risk category

The paper correlates the climatological parameters of temperature and precipitation (data series available for 50 years) to assess associated hazard risks. The total resulting risk is estimated for each county in Romania (as an application case). Based on the mapped risk, an insurance premium is calculated on a per capita basis. Conclusions are drawn related to monitoring this type of risk and introducing insurance coverage. The risk assessment methodology for risk mapping is employed for four types of CC event that induce risks (floods, drought, snow, and freezing). The risks are determined from the correlation of climatological parameters and not from the simple analysis of specific events, e.g., the occurrence of floods. The main message is that mapping risk at the distributed county level may provide useful information for risk management policies such as insurance for CC event-induced risks.

4.4 Risk assessment frequency/probability measures

Risk is given by the product of the frequency (probability) of an event and the damage produced. The CC events considered here are generated by the combined effects of temperature and precipitation. To assess probabilities of temperatures being under/above their given values and of the precipitation being under/above their determined limits, data from NAM are used, containing monthly average data for a 50-year time series of temperatures and precipitation for each county in Romania. The values are spatially averaged for each county from the data recorded by the monitoring networks of NAM. As an example, the data series for temperature and precipitation for Arges County is provided below.

From these sets of basic data, several probability distributions are calculated that best fit the temperature and precipitation time series in each county. To make basic calculations, only the normal distributions determined for each parameter and county have been considered for the respective average and standard deviation values. Using a large amount of data also helps determine probability distribution functions that integrate short-term fluctuations in parameter values. The exercise performed here is intended as a basic example that can be (and is recommended to be) extended to short-term fluctuations, extreme values distributions, combined effects, e.g., of melting snow and rain, expected damage distributions for each county, and other characteristics of interest.

Having these distributions based on real data allows probabilities to be calculated for each parameter and their combinations. Thus, this exercise has two options for temperature and two for precipitation, resulting in four events as depicted in Fig. 4.3. The probabilities are calculated for the year following the time series data—each new year adds new data that may change the distribution parameters after a new distribution calculation is made.

The event tree at the county level is presented, chosen at random for Arges County. The normal distributions with their moment values are used in the assessment of each county (Fig. 4.4).

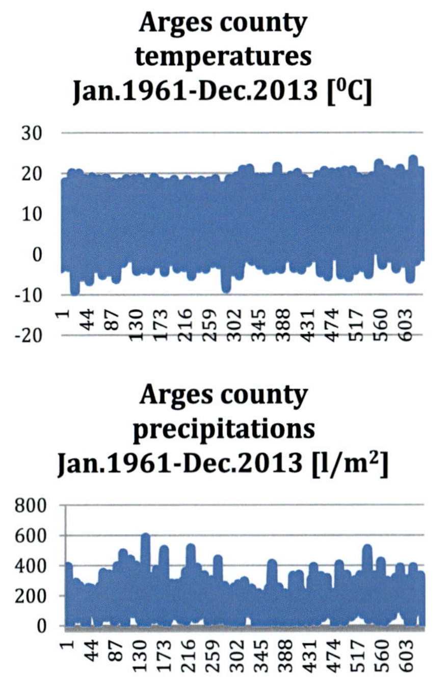

Figure 4.3 Climatologic parameters for Arges County.

The event tree can determine the frequency of occurrence of floods, drought, snow, and freezing from a combination of climatologic parameters (i.e., temperature and precipitation). The frequency of each event is usually determined from the analysis of its occurrence and not from a combination of climatologic parameters. This approach has the advantage of using meteorological data recording networks and making a closer connection with the probabilities used in the event tree.

To understand the event tree, note that the limit between low and high temperatures is zero degrees Celsius (this is the difference between rain and snow precipitation), i.e., with negative temperatures there is no flooding but only heavy snow when precipitation is high and freezing when precipitation is low; conversely, when the temperature is above zero, high precipitation produces floods while low precipitation produces drought. More complex scenarios may be considered where melting snow combines with rain at above-freezing temperatures, thus acting as an aggravating element. The event tree in such cases can be extended to accommodate more options. This report only serves as an example of a first-order risk mapping and is not an extended risk evaluation.

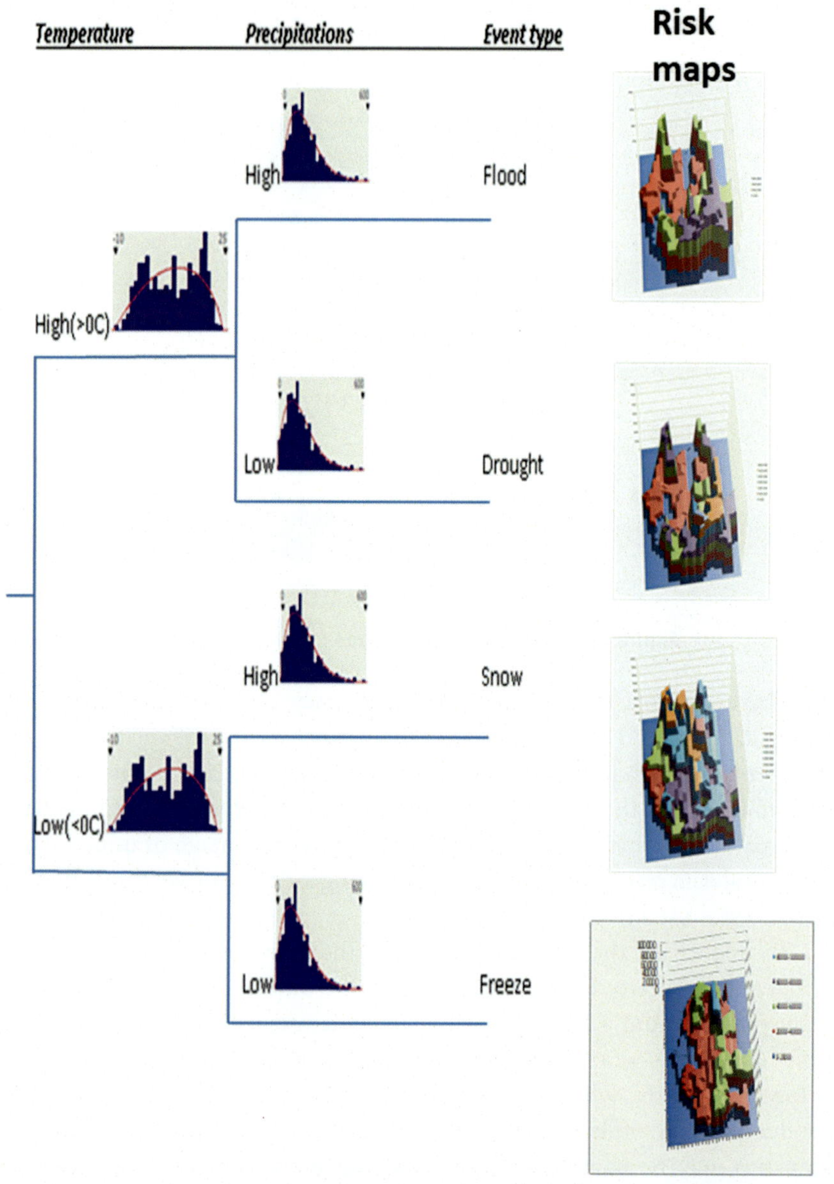

Figure 4.4 Event tree for climate change events.

4.5 Assessing damage

The other component of risk is damage associated with each type of event and for each county. Here the data are not as diversified as with monitored climatologic parameters. The basic sources are the UNSDR database for Romania (whole country average values for floods and droughts), data on interventions of the EU Solidarity Fund (damage values or snow and freezing), data on floods provided by a World Bank (WB) project on hazard risks in Romania associated with floods, and data from a WB study on CC risk in Europe and Central Asia (ECA) that provides general estimations. Based on these data, average damage per type of event has been determined and was considered the same for each county. Since this analysis is simply opening the way to more elaborate and systematic determinations, this approximation is used to determine the county risk values per type of CC event considered. The sum of these risks is presented in the section on combined risks. The values considered (see Annex 3) for the damage per type of event are the following:

(1) Flood—328 million US$/event

(2) Drought—250 million US$/event

(3) Snow—300 million US$/event

(4) Freeze—300 million US$/event

These values are just for orientation and are considered equal for each county. A thorough program to determine the damage for each of the above events by county must be implemented for a more detailed risk assessment.

NOTE: the type of analysis done here is not an extreme value analysis for the resulting four events. The purpose of this exercise is to assess the correlation between the climatologic parameters (temperature and precipitation) and the potential outcome events described above using an event tree method. The peak-over-threshold method is not used here to determine extreme values associated with a given distribution to determine the exceedance probability curves. Also given is the basic assumption that damage is the same for all counties; it is not the purpose of this analysis to determine a farmer curve. The values for the data series are fitted to a normal distribution so that probabilities can be calculated for various parameter value scenarios (such as negative or positive temperatures). The damage is considered the same for all counties (and further research to determine the distributed damage is strongly recommended). The resulting risk is measured for the year after the data series and every new real data point that adds to the series requires a new determination of distributions to assess probabilities. Moreover, damage is changed by investments allocated for mitigation measures such as, e.g., flood protection systems, retention areas, water releases from upstream dams, etc. These changes assess new values for damage (hence for risk) that determine new investment allocation. The process is proposed here without further details as just a basic exercise.

4.6 Climate change risk maps

The risks of the above CC events are calculated as the product of the probability and the damage. The resulting maps for each risk at the county level are presented below. The presentations are based on two types of maps: (1) one in 3D where it is easy to spot at a glance the areas of high risk and (2) in 2D where the view of each county is better defined. The data are calculated in Excel, which is also the source of the 3D maps. In each map (of the four types of hazard risks described above) there are four (e.g., for flood risk) or more (e.g., for snow risk) ranges of risk values to facilitate, in each case, a better perception of the main differences; various levels of detailed ranges (e.g., five for the total risks, or several for the gas grid risks) may be defined for the 3D maps as allowed by Excel.

4.6.1 Flood risk map

The first map is the one of flood risk. Here the Northwest and the East and South parts of the country are the most affected along with the counties positioned along the Danube. The areas affected by the 2005 floods were found in the same area. It should be mentioned, though, that there is an effect from precipitation outside Romania that is brought in by the Danube that is not considered here (Fig. 4.5).

4.6.2 Drought risk map

Drought risk has a similar pattern to flood risk, which seems right because river basins produce floods in some years but in other years are subject to lower precipitation or higher evaporation and hence to drought. Obviously, the detailed probability values and damage are different in the two cases. Another observation is that the investments required to protect against flood effects are not the same as those to manage the effects of drought. Areas with high drought risk, e.g., the southeast part of the country, also contribute substantially to agricultural production, whereas the central and western parts of the country are spared the impacts of drought. This situation may provide indications of how to allocate investments for developing irrigation projects to effectively mitigate drought effects (Fig. 4.6).

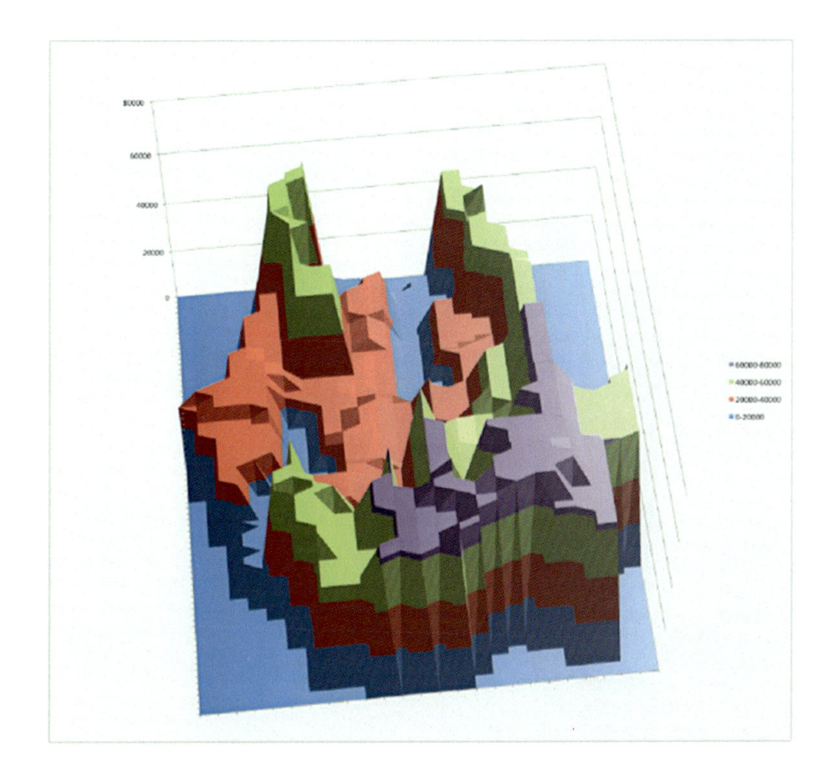

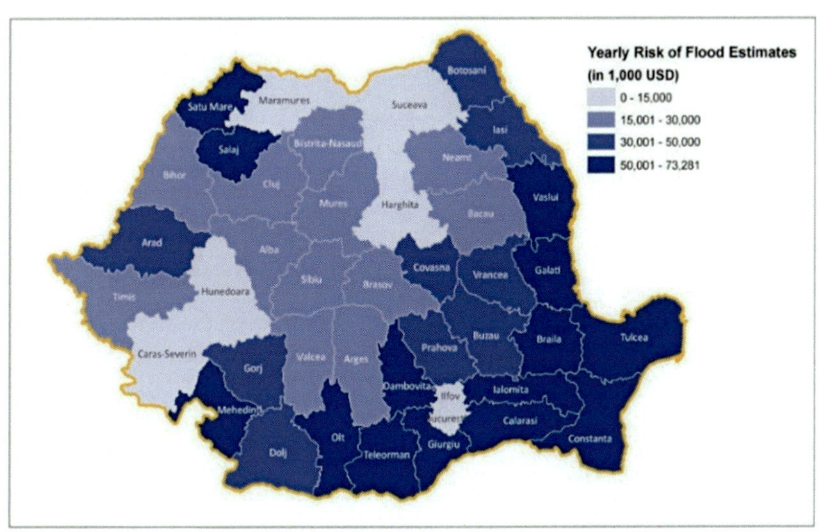

Figure 4.5 Flood risk (1000 USD per year). *Based on authors' calculations.*

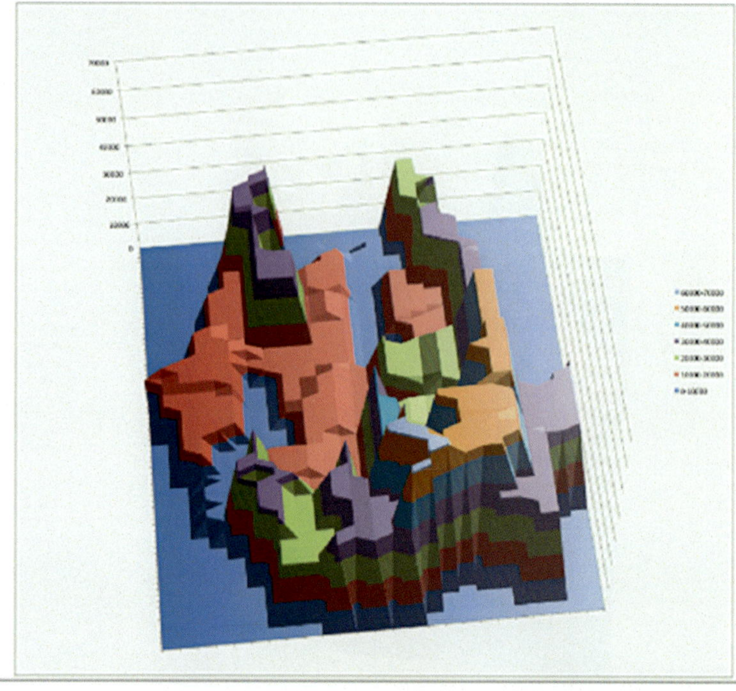

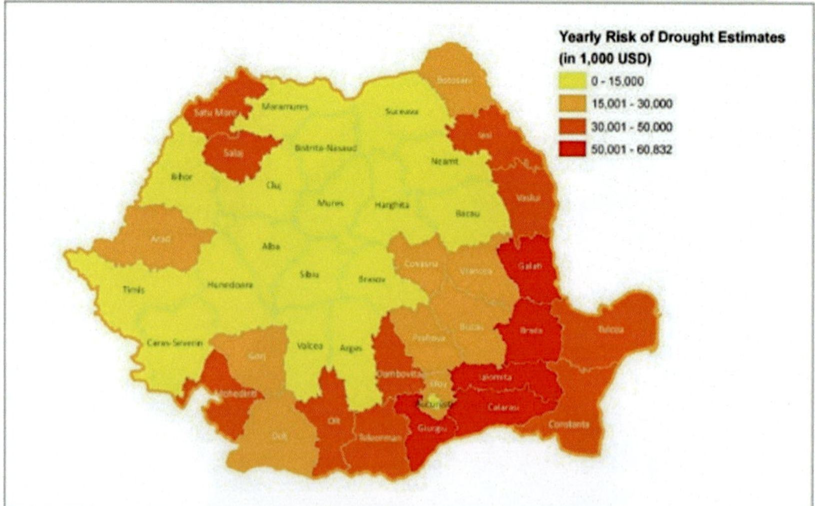

Figure 4.6 Drought risk (1000 USD per year). *Based on authors' calculations.*

4.6.3 Snow risk map

The risk associated with heavy snowfall presents a different pattern than the previous events for counties in Romania. This time, the east and central areas are most affected, while the seaside and southwest (which has a Mediterranean climate) are less prone to heavy snow. The mountain areas show high snow risk values, and the east and southeast plains regions also have a sizable risk. This situation provides indications as to how to distribute snow fighting to the counties of Romania, how to prepare for situations of interrupted road and rail transport due to heavy snow, and how to prepare for interruptions in electrical energy and gas supply network operations (Fig. 4.7).

4.6.4 Freeze risk

Freeze risk has a pattern similar to that of snow. Regions with higher freezing conditions include the region of Covasna County, which for a long time has been called the pole of frost in Romania and is seen as the highest among the counties. The northwest is also colder, whereas the southwest and the seaside areas are spared from freezing. As in the case of snow, the east and southeast plains are exposed to frost, and since this case is associated with low precipitation, there is an impact on agricultural production that must be considered. Moreover, in low temperature conditions, the mechanical failure of, e.g., gas network pipelines, has different probability values associated with the behavior of materials at low temperature—this is considered in the calculation below for the risk of gas network failure (Fig. 4.8).

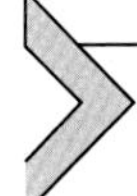

4.7 Mapping tool and combined climate change effect risks

To map risks associated with each county, a tool was built in Excel where the risk value for each county is inserted into the cases associated with the surface of said county. Thus, the surface of Romania is covered by groups of cases that map each county. The representation of a 3D figure in Excel provides a risk map for Romania presenting each county surface with a given risk height. Maps made in this way have the advantage that the risk values result from elaborate calculations made possible by Excel. Combining several such maps with varying parameters, a risk dynamics analysis can be represented. As an example, the total CC event risk map—summing the separate risks presented above—is provided below (Fig. 4.9).

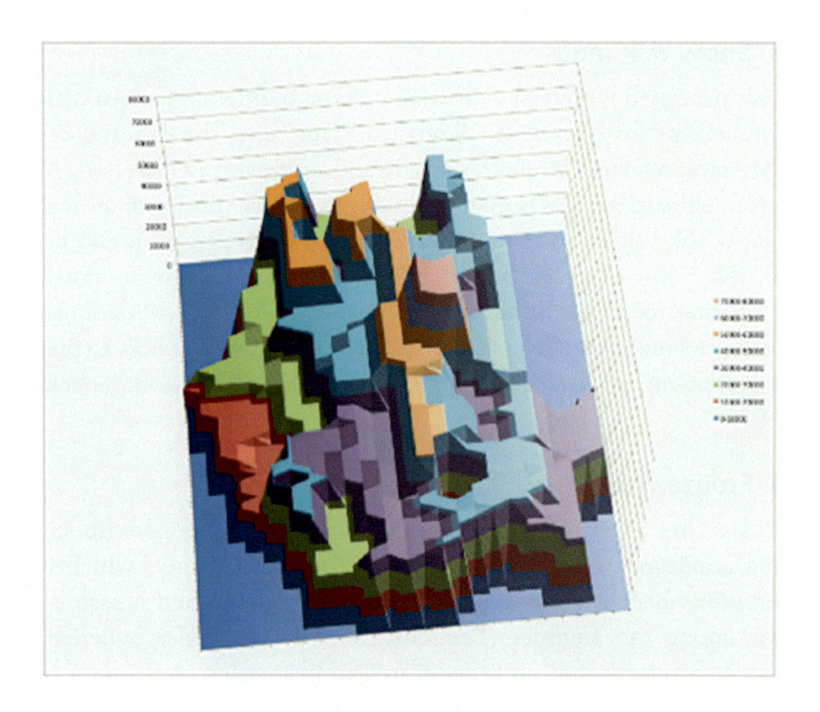

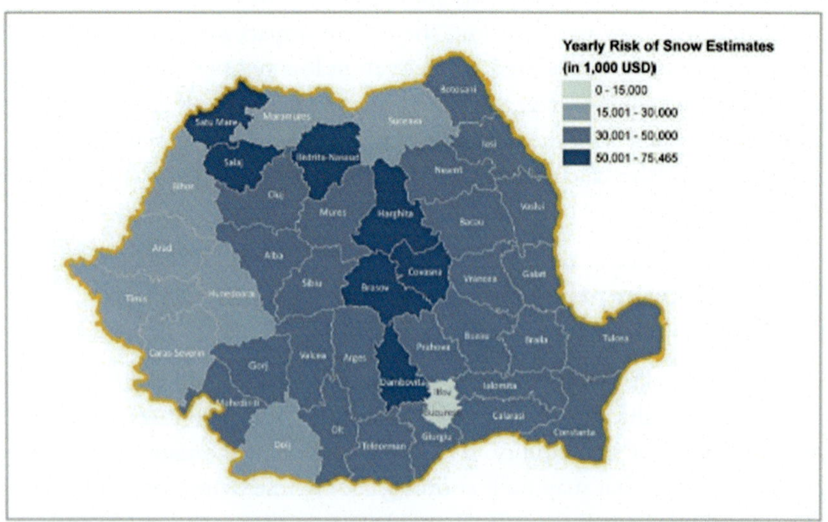

Figure 4.7 Snow risk (1000 USD per year). *Based on authors' calculations.*

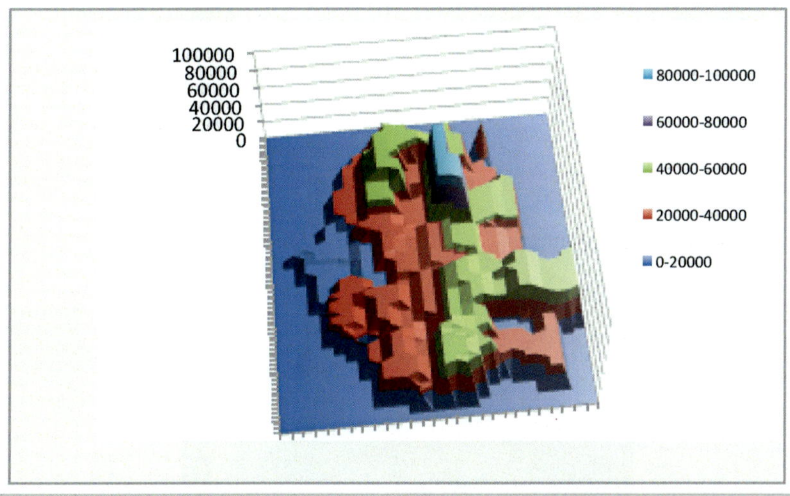

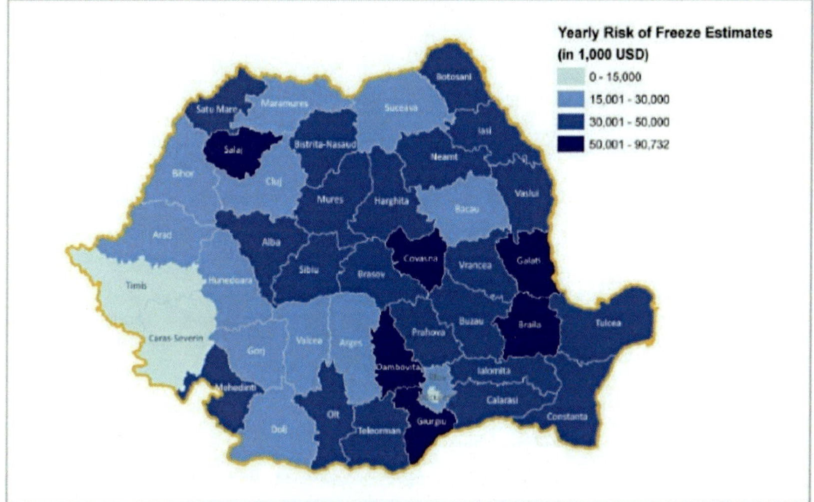

Figure 4.8 Freeze risk (1000 USD per year). *Based on authors' calculations.*

These risk values are considered in the analysis that follows and used to draw conclusions about the support information for a CC event risk insurance policy setup.

A good example of combined risk effects is the assessment of the exposure to CC event risks in each county. Exposure is determined based on the GDP of each county and the risk values in US$ determined above. Performing the calculations for the whole country results in a risk/GDP ratio, expressed as a percentage of GDP, that amounts to slightly over 3% of

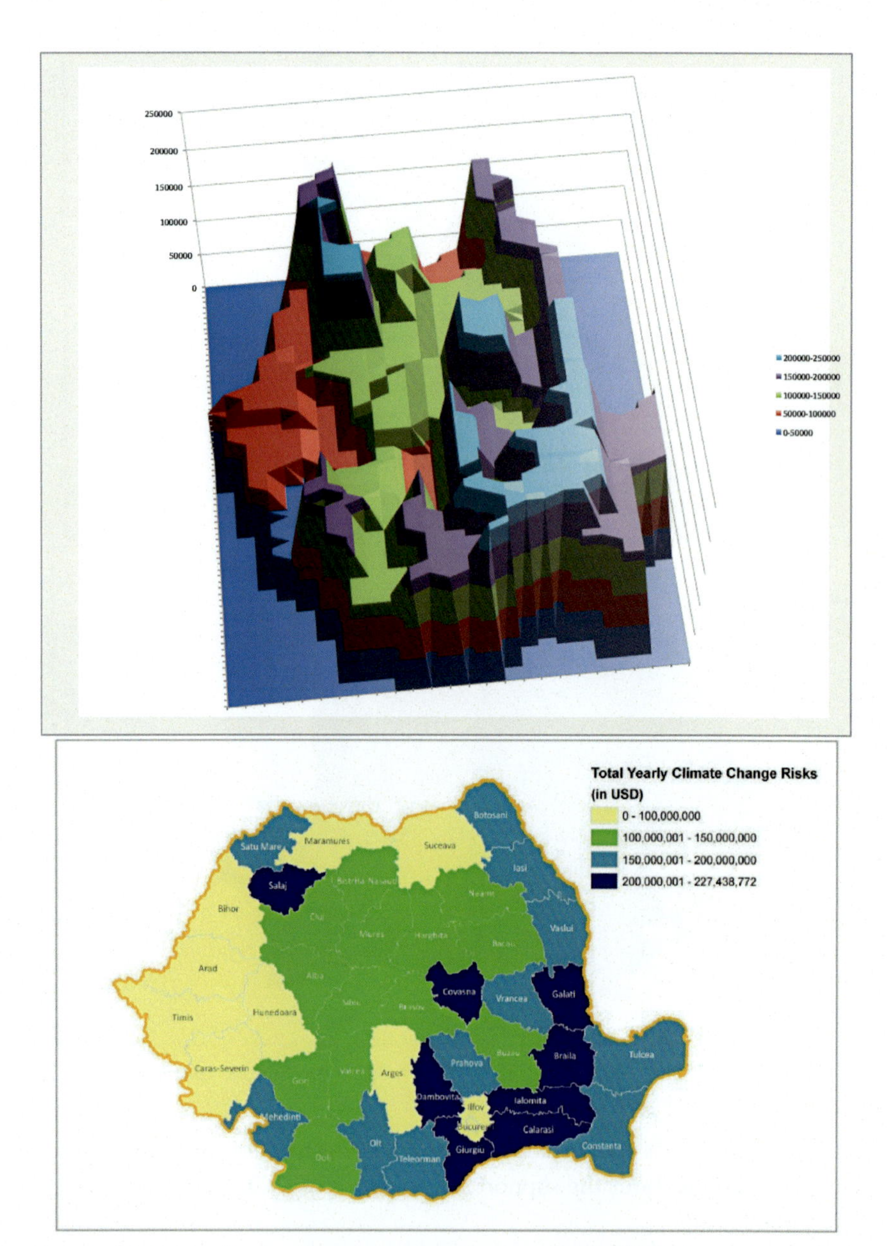

Figure 4.9 Total climate change event risk map (1000 US$). *Based on authors' calculations.*

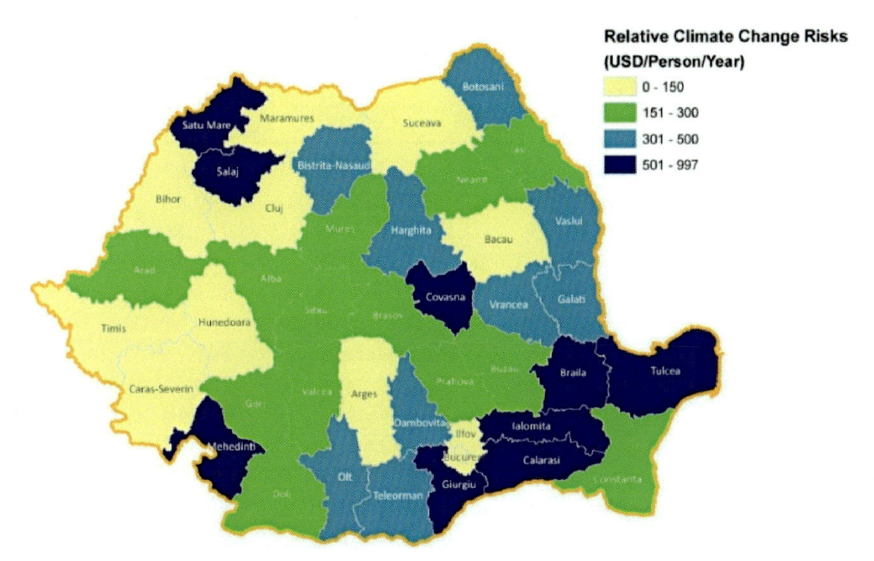

Figure 4.10 Two-dimensional map representation of the data. *Based on authors'calculations.*

GDP. The values resulting from this risk evaluation method (that do not consider all hazards) are comparable to the ones determined for Romania in a WB report on hazard risks in ECA countries (John Pollner et al., 2010).

Further on, considering the risk value at each county level divided by the population of each county, one may calculate per capita exposure as the basis for a better distribution of insurance coverage for each county (Fig. 4.10). Table 4.1 summarizes these results.

4.8 Population at risk and economic impacts

The population at risk is different for each type of risk. The differences stem from geographical position, poverty level, access to critical infrastructure, etc. Thus, vulnerability to risk is not the same for the various population categories. This situation diversifies the policies aimed at reducing these vulnerabilities. Devising adaptation and mitigation measures must account for the impacts associated with the entire range of hazards.

For example, flood risk has a greater impact on residential areas closer to river basins' flood-exposed areas. In contrast, in the case of flash floods, populations are more exposed in areas where torrents may form. In the case of snow, the most exposed population are in places where means of snow protection are missing, critical infrastructure (e.g., roads or power grids) is

Table 4.1 Distribution of risk per capita—counties in alphabetical order (Bucharest is set to 0 such that only Ilfov County, where the city is located, is considered.).

County	Risk/cap US$	County	Risk/cap US$
Bucuresti	0	Harghita	328.07
Alba	284.22	Hunedoara	140.63
Arad	196.9	Ialomita	726.46
Arges	138.01	Iasi	204.44
Bacau	138.87	Ilfov	111.41
Bihor	140.54	Maramures	138.51
Bistrita Nasaud	454.83	Mehedinti	542.71
Botosani	342.11	Mures	188.71
Braila	584.28	Neamt	193.19
Brasov	215.93	Olt	351.47
Buzau	270.01	Prahova	196.16
Calarasi	669.42	Salaj	852.13
Caras Severin	145.67	Satu Mare	518.91
Cluj	146.23	Sibiu	294.8
Constanta	231	Suceava	85.53
Covasna	996.85	Teleorman	395
Dambovita	381.84	Timis	111.87
Dolj	165.01	Tulcea	610.18
Galati	347.21	Valcea	256.91
Giurgiu	767.9	Vaslui	378.84
Gorj	279.39	Vrancea	424.98

weaker, or the frequency of heavy snow is higher. The case of freezing affects populations in regions with a low temperature microclimate and where winter conditions are more severe. In addition, very low temperatures may influence flow within the gas grid, or the creation of ice on high-voltage power transmission lines, with damaging economic effects.

A typical example of a population at risk in flood-prone areas of Romania is given in the map below (Fig. 4.11).

The rise in the frequency and consequences of hazards in Romania has led to increased awareness by the administration related to implementating appropriate measures. The necessary legal framework is enhanced by the EU legislation transposed into national laws.

4.8.1 The way forward, a synthesis for decision-makers

This last section is intended as a synthesis of recommendations related to all the monitoring and risk mapping systems described above.

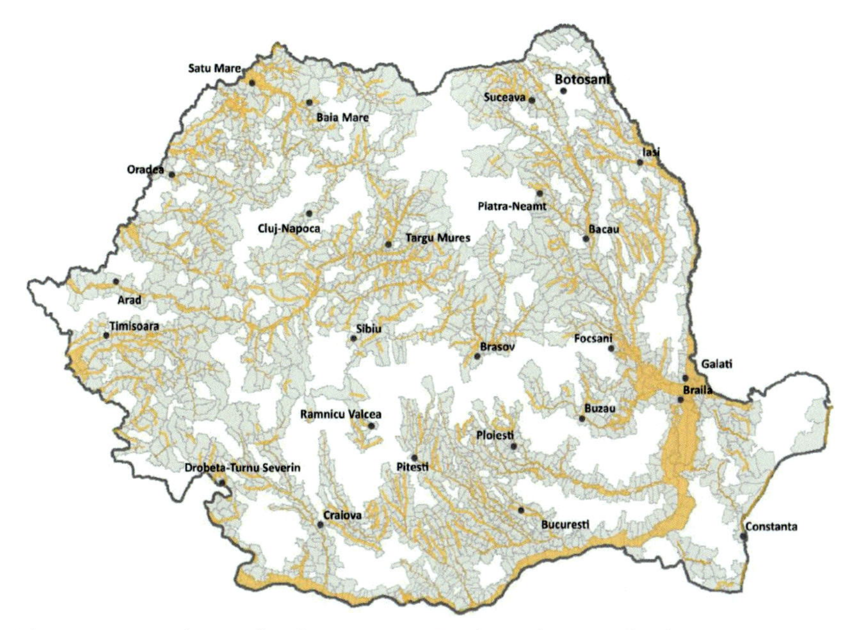

Figure 4.11 Localities in flood prone areas. Total population in flood prone areas: 16.3 million (79% of the total population). *Based on Purica, I., 2014. Report on Hydrological Basins Distribution of Population. ANAR, Romania.*

4.8.2 Authority of the General Inspectorate for Emergency Situtions

On a general level, the institutional setup in Romania to monitor CC events has one important flow—the General Inspectorate for Emergency Situations (GIES)—but it does not have the status of a ministry. For example, in the case of EU Directive 2008/114/CE implementation, this status needed the foundation of a dedicated center for critical infrastructure within the Ministry of the Interior. This center has the authority to cooperate with entities from other ministries that coordinate critical infrastructure and with the companies involved.

As CC events start to manifest themselves with more intensity, it is recommended that GIES be given interministerial status, possibly depending directly on the prime minister or the general secretariat of the government to have the appropriate level of authority for the complex interventions required by the manifest risks of CC events.

4.8.3 Investments

The evolution of CC events requires not only keeping Romania's existing monitoring systems in operation at nominal capabilities but also upgrading

EU requirements and even going beyond this to modern monitoring systems that allow real-time dynamic surveillance of various parameters with both a predictive purpose and use as reactive action feedback. The investments needed range from 10 to over 100 million euros. It is very important to put dedicated investments in the annual budget every year and to have a policy on CC events mitigation and adaptation investments that includes monitoring systems use, maintenance, and development as a permanent position. The use of EU ESIF in combination with other sources of finance is recommended. Moreover, this may also help to define components related to CC to be 20% in every investment project as per EU requirements. The allocation of investments for mitigation measures may be done based on the exposure and changed to the remaining exposed areas as the change in the exposure diminishes from implemented investment projects. This may also provide a reason to better monitor the ongoing investment projects.

4.8.4 Communication

Both predictive and reactive actions for CC event risk monitoring and mitigation are based on a reliable and extended communication system. It is strongly recommended that present communication capabilities be extended. CC events are increasingly perceived as strategic threats that require complex interventions, and monitoring and communication systems should be regarded as strategic elements and even be included in Romania's defense strategy.

4.8.5 Insurance

The EU recommends implementing insurance policies for CC events' induced risks and has started definite measures in this sense. The mapping of risk shown above provides a basis for exploring the implementation of CC insurance in Romania. It is recommended to continue the analysis of the measures to have in place for an appropriate system that may also contribute to having a fund to finance various mitigation and adaptation measures. Considering a given insurance premium based on, say, 0.5% of the risk, one may see from the risk exposure in Table 4.2 that the premium is affordable for the average-income population. This opens ways to implement the appropriate legislation for a hazard risk insurance policy similar to auto or home insurance policies already in operation.

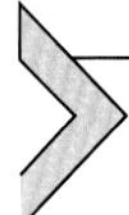

4.9 Setting the basis for a climate change event risk insurance policy

The information on combined risk effects is used to assess the exposure to CC event risks for persons in each county. The exposure is determined based on the population of each county and on the values of risk measured in US$ determined above.

The values resulting from this method of risk evaluation are comparable to the ones determined, for Romania, in a WB report on hazard risks in ECA countries.

Moreover, considering the value of each county risk level, divided by the population of each county, one may have the exposure on a per capita basis. Considering an insurance premium of 0.5% of the exposure (slightly larger than typical insurance premiums for other types of risk) leads to the foundation for a distribution of insurance coverage for each county. Table 4.2 summarizes this result.

Table 4.2 Distribution of risk premium per capita—counties in alphabetical order. (Bucharest is set to 0 on purpose such that only Ilfov County, where the city is located, is considered.).

County	Premium risk/cap US$	County	Premium risk/cap US$
Bucuresti	0	Harghita	19.68
Alba	17.05	Hunedoara	8.44
Arad	11.81	Ialomita	43.59
Arges	8.28	Iasi	12.27
Bacau	8.33	Ilfov	6.68
Bihor	8.43	Maramures	8.31
Bistrita Nasaud	27.29	Mehedinti	32.56
Botosani	20.53	Mures	11.32
Braila	35.06	Neamt	11.59
Brasov	12.96	Olt	21.09
Buzau	16.20	Prahova	11.77
Calarasi	40.17	Salaj	51.13
Caras Severin	8.74	Satu Mare	31.13
Cluj	8.77	Sibiu	17.69
Constanta	13.86	Suceava	5.13
Covasna	59.81	Teleorman	23.70
Dambovita	22.91	Timis	6.71
Dolj	9.90	Tulcea	36.61
Galati	20.83	Valcea	15.41
Giurgiu	46.07	Vaslui	22.73
Gorj	16.76	Vrancea	25.50

4.10 Decisions based on risk

The mapping of CC risks is not a one-time exercise but is subject to a Pareto analysis. As investments are allocated to the more exposed counties, the potential damage in these counties is reduced and other counties are becoming more exposed. A new risk mapping will put those in the first position for investment allocation and so on.

Moreover, the risk exposure of the counties may be correlated with various other economic and/or social parameters such as the contribution to the total GDP or the exposure of the poor population, etc. This may make the decisions on socio-economic policies and measures more coherent and provide a clear measure of the impact of said decisions.

4.10.1 Further actions

This paper opens the way for more activities related to monitoring and using the data for the assessment of CC events' induced risks. The application provided a methodology that needs to be diversified, e.g., by determining the distributed potential damage values at the level of each county. This is important because in the case of CC events, the probabilities are not directly controllable (but only indirectly through greenhouse gas emissions reduction). Thus, the mitigation and adaptation actions are acting on the damage values that are diminished through appropriate investments. A WB project for Romania on hazard risks that concentrated on flood risk mitigation investments provided good information that needs to be extended to other CC events.

Further on we give a case example of applying the distributed risk to assess the impact on the gas grids of Italy for earthquakes and landslides and Romania for the CC risks presented above.

4.11 Hazard risks and their impact on critical infrastructure

4.11.1 Case analysis—natural gas networks of Italy and Romania

The interconnection of critical infrastructure represents one of the pillars of the EU Energy and CC strategy for the horizon of 2030. The risks associated with these networks should be analyzed based on the geographical distribution of each network by contrast to local objectives such as nuclear power plants or dams. Based on distributed hazard risks evaluation done for

Italy's regions in case of seismic and landslide and for Romania's counties in case of flood, drought, snow, and freezing and on the risks of mechanical failure with gas escape and ignition, the risk map is determined for each country measured in probable deaths per million inhabitants. These results may provide the needed information for optimizing the allocation of mitigation means and for implementing efficient insurance policies.

4.11.1.1 Introduction

4.11.1.1.1 Quantification of risk

The quantification of risk has always been an intensely debated subject. Early reports such as WASH-1400 and "The Canvey Island Reports" have tried to assess valid scenarios on which to base the various physical and statistical analysis headings to determine frequencies of accidents and intensities of their consequences.

Since there is no general method for establishing all the consequences to be considered as negative and which of them, if not all, should be analyzed, there is no way to be certain that a complete analysis has been performed.

Moreover, when such an analysis is done on a complex system, such as the NG system in Italy and Romania an embedded structure is encountered having various levels of complexity at different scales. When analyzing the data, one has to disaggregate at various scales and identify the intercorrelations both horizontally and vertically within the structure.

Inevitably the uncertainty is rather high in relation to some data, the time horizons of different papers and statistics do not completely overlap, and there is not a consistent way of reporting the data.

4.11.1.1.2 Data preparation

When considering the data to quantify risk at the level of the Italian and the Romanian natural gas system we are faced with a time evolution and space distribution.

The space disaggregation of data is decided by the intersection of the sets of available consistent data, which for whole countries is given by the regional distribution.

From this point of view, we distinguish among three types of data: (1) data reported on a regional distribution basis that provide total certitude at this level (e.g., population, surface, NG consumption); (2) data reported in absolute values aggregated over the whole country that we had to consider as uniformly distributed (like the probability of gas ignition); and (3) data reported in specific values in correlation with other data whose distribution we know. Based on the known distribution we may generate a distribution of the unknown data (with a certain level of incertitude, though), e.g., the frequency of gas pipe rupture may be generated for each region based on the number of pipe kilometers in each region and the frequency of rupture per kilometer that is reported in various papers.

The data storage and manipulation have been performed in the "Excel" computer program environment.

4.11.1.1.3 Literature review

The problem of determining the risks of networks exposed to hazards related to earthquakes, landslides, or CC-induced events (floods, drought, snow, and freezing) has started to be explored in Italy by ENEA in the early 1990s (Purica, 1991) and has continued to be expanded both by the EU (Eu Commission, 2007) and by the IBRD, Pollner et al. (2010) (later on as the impacts of CC events had become more visible) and IPCC in 2013. The assessments of risk, though, were done mostly for floods (based on the occurrence of these specific events), and the disaggregation was essentially regional. No connection was seen for CC parameters, such as temperature and precipitation. The papers by Purica (2010, 2014) extend the methodology for assessing the risk of critical networks, such as gas networks, for a given country with regional distribution and based on assessing risks of hazard events that result from combinations of data series for temperature and precipitation at each regional level and over long periods (50 years). So the combination of earthquake and landslide risks for Italy and that described above for Romania is associated with a mechanical defect and the population at risk to determine a risk map for the critical network of gas within each country.

4.11.1.1.4 Logical model and risk mapping

From the standpoint of risk, two types of risk may lead to potential deaths and disabilities.

The first type stems from the fact that methane is primarily distributed through a network of pipes with a practically uniform distribution over the total surface of each of Italy's regions and Romania's counties.

Of course, we stick to regional disaggregation; if we go below that, we introduce completely different distributions of data that must account for big cities, industrial platforms, power plants, and so forth. If an assessment is made for a specific area, the distributions mentioned above must be considered, but the methodology presented below will be the same.

On average for all of Italy in 1986, each square kilometer of surface area served by the gas network (i.e., Sardegna excluded) had 76.2 m of principal transport pipe and 334.4 m of distribution pipes. Put another way, we may say that each kilometer of pipe corresponds to a surface area of 2.4 km^2. In other words, there is 1 km of methane pipe in each square measuring 2.4 km^2 with 1.56 km sides. Taking the population density of 190 inhabitants/km^2, we can determine that each 2.4 km^2 has 463 inhabitants on average.

We did all this averaging just to point out that gas escape has a sensible probability of ignition and that people may be affected.

Based on the data described previously, we may consider that the surface of every region has a percentage that shows ground movements either as ground instabilities or as seismic movements. Both ground instabilities and seismic movements are characterized by a surface affected in every region and by an intensity of movements expressed as the number of movements recorded per region per year to represent seismic intensity on the Mercalli scale.

In the case of ground instability, we have expressed both the number of areas per 100 km^2 and the intensity of movements as regional percentages from a country total. Considering the seismic case, there are three levels of seismic intensity surfaces, so we summed the surfaces and expressed them for each region as a percentage, and we also calculated an intensity as a surface-ponderated regional index expressed as a percentage of the country total. After this normalization, considering the ground movements, we may distinguish among four types of surfaces: (1) having both ground instability and seismic movements; (2) with only ground instability; (3) with only seismic movements; and (4) without instabilities.

If we express this in Boolean logic, putting g-ground and s-seismic percents of unstable surfaces, we find that for each region, r, of surface, Sr, the unstable surface is given by $Sr\ (g\ s+g\ (1-s)+(1-g)s)$, while the stable surface is $Sr\ (1-g)\ (1-s)$.

Considering that methane pipe with a certain kilometer length is distributed on the surface of each region, we may assess that some pipe will pass through ground movement-affected surfaces. So the distribution of the frequency of ground movement-caused accidents is not the same for all the lengths of pipe within a given region, and since we assumed a uniform distribution of pipes over each region, the length of pipe will be ponderated by the same ground movement coefficients as the surfaces of those regions.

If we look now at the frequencies of gas escape incidents, we see a rather sharp behavior limit established by the 16″ pipe diameter. A higher frequency of incident exists below 16" but with a lower probability of ignition, while for diameters over 16″, the incident frequency is small but with a high ignition probability due to the large gas masses involved.

Based on the above comments we may separate several categories of causes, which have specific incident frequencies: (1) causes that are not sensitive to diameter or ground movement like construction/materials; corrosion; other causes; (2) causes that are sensitive to ground movement. They increase the frequency of accident for the pipe length affected by ground movement; (3) causes that are sensitive to diameter variation as hot tapping and external interference, which apply respectively to pipe lengths having diameters lower and/or greater than 16″.

Representing the above into a logical tree for each type of pipe damage: *pinhole* (p)- diameter of defect smaller or equal to 20 mm; *hole* (h)- diameter of defect greater than 20 mm; *rupture* (r)- diameter of defect greater than pipe radius; we obtain Fig. 4.12.

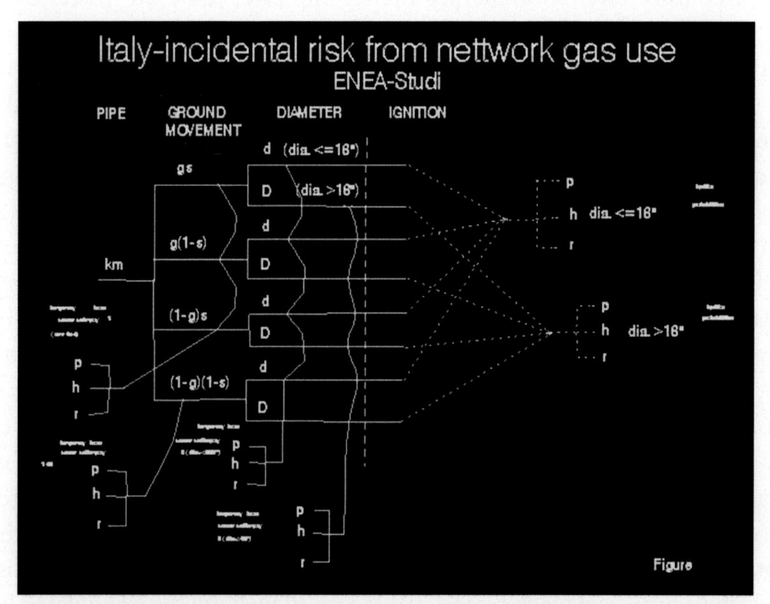

Figure 4.12 Event tree for risk evaluation. *From Purica, I., 1991. La dinamica dei Rischi nel Sistema Gas Naturale in Italia, Report in Program VESE. ENEA, Roma.*

Calculating the branches of the tree and summing for each defect type we obtain.

Defect type	Formula (L-pipe length)
p	(L) ((gs+g(1−s)+(1−g)s) * 0.24E-3+(1−g) (1−s) * 0.23E-3) (d * 0.5E-3+D * 0.02E-3) * 1.6E-2
h	(L) ((gs+g(1−s)+(1−g)s) * 0.062E-3+(1−g) (1−s) * 0.043E-3) (d * 0.97E-3+D * 0.05E-3) * 2.7E-2
r	(L) ((gs+g(1−s)+(1−g)s) * 0.014E-3+(1−g) (1−s) * 0.012E-3) (d * 0.47E-3 * 4.9E-2+D * 0.07E-3 * 35.3E-2)

Summing on all defect types to obtain the total ignition accident probability we have:

(L) ((gs+g(1−s)+(1−g)s) (dm+Dn)+(1−g) (1−s) (du+Dv))

where m = 3.78E-9; n = 5.06E-10.u = 3.24E-9; v = 4.28E-10.and d = 69.6%; D = 30.4% only for transport pipes.d = 94.02%; D = 5.98% for transport and distribution pipes.

Until now, we have calculated the values of the frequency of gas releases and release followed by ignition. Since risk is defined as the frequency of an adverse event multiplied by its consequences, we shall now take a look at the consequences of the types of incidents our analysis involved. The consequences of accidents from using the NG network are analyzed in Purica (1991), and we do not repeat them here.

The calculations of risk expressed as probable deaths from the use of network gas in every region of Italy, taking into account the specific data for each of them, are presented in Fig. 4.13.

We must finally mention, for comparison, that the mortality index for the Italian electrical grid distribution system was approximately 4 (deaths/E6.inhabitants) in 1983. This indicates that the NG network is as safe as the electric grid if compared with a total Italian all-energy sources incident mortality index of 13 (deaths/E6.inhabitants) in 1983.

Another risk assessment case mapped for this study relates to the influence of the CC events mentioned above on critical infrastructure in particular the gas network of Romania. The assessment is done as an example of critical infrastructure that responds to EU Directive 2008/114/CE. The analysis is similar to the one done for Italy, replacing the values of the four combined seismic and franuosity risks with the four types of hazard risks

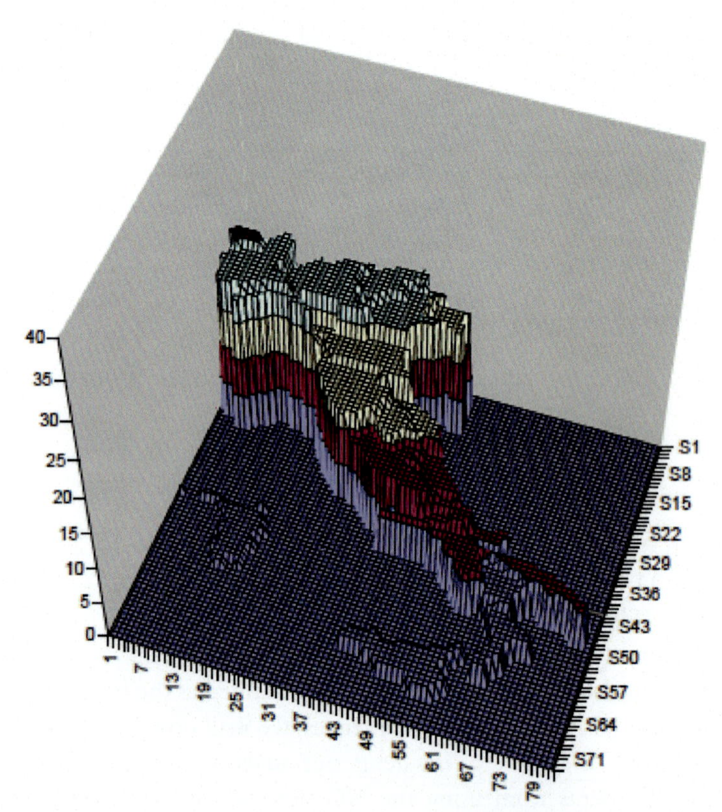

Figure 4.13 Natural gas risk in Italy (probable deaths/million inhabitants) (Purica, 1991).

(flood, drought, snow, and freezing) as evaluated in a previous paper (Purica ESPERA, 2014).

The assessment starts with the probabilities determined for each event and each county. Then it considers the number of km of gas network in each county (given by the National Institute of Statistics of Romania). The event probabilities are combined with the mechanical failure probabilities of gas pipelines, based on a more elaborate event tree (see Purica 1991, 2010) and the above calculation for Italy. The combination of these two types of probabilities results in the gas escape probability due to CC events followed by mechanical failure.

Considering the population at risk as the one supplied by the gas network and the impact, in probable deaths per gas escape event, from gas grid accidents' estimations, the risk is determined as measured in potential deaths per 1000 inhabitants, from gas escape events, in each county. The resulting map of this type of risk is presented below (Fig. 4.14).

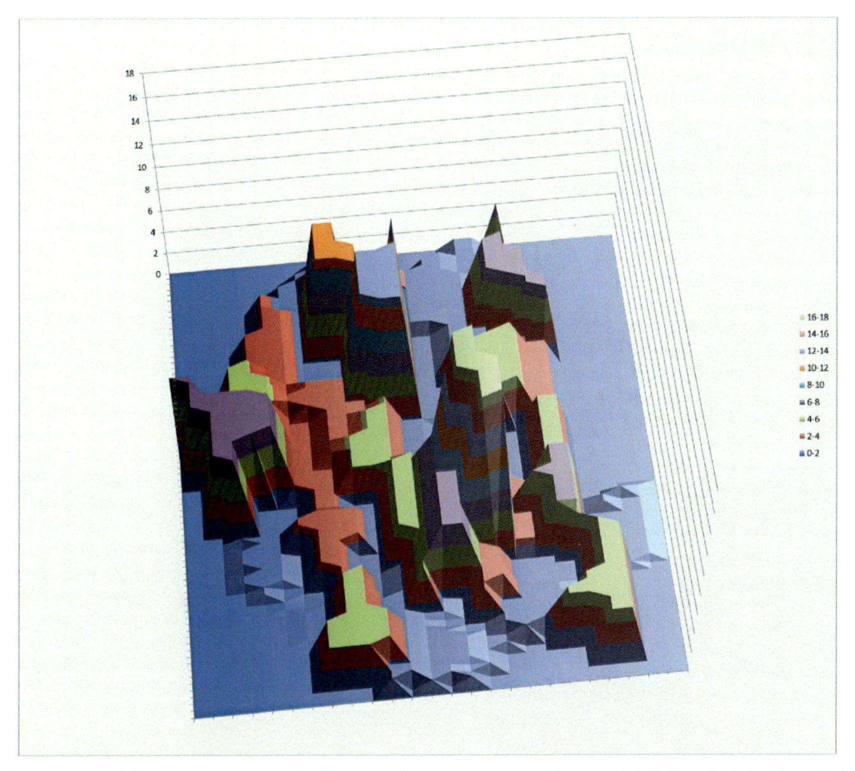

Figure 4.14 Romania gas grid climate change and mechanical risk (probable deaths/million inhabitants). Based on author's calculations

The map shows the areas where the gas grid is more developed having a higher risk. The probabilities of mechanical failure are based on estimations done for similar material pipelines in Italy—Romania does not have at present a consistent activity of determining and reporting these values.

4.12 Conclusions

The methodology developed here for the mapping of network distributed risks allows among others the introduction of a policy to optimally allocate mitigation means among the regions/counties of each country to minimize the intervention time in case of accident and to devise insurance policies better adapted to this type of risk coverage.

Moreover, with appropriate data, the method may be extended to other types of critical infrastructure risk mapping, providing the possibility to better face the requirements of the EU energy and CC 2030 strategy.

Annex 4.1

The evolution of temperatures average and standard deviation by 10-year periods, starting in 1961, for each county.

Average values.

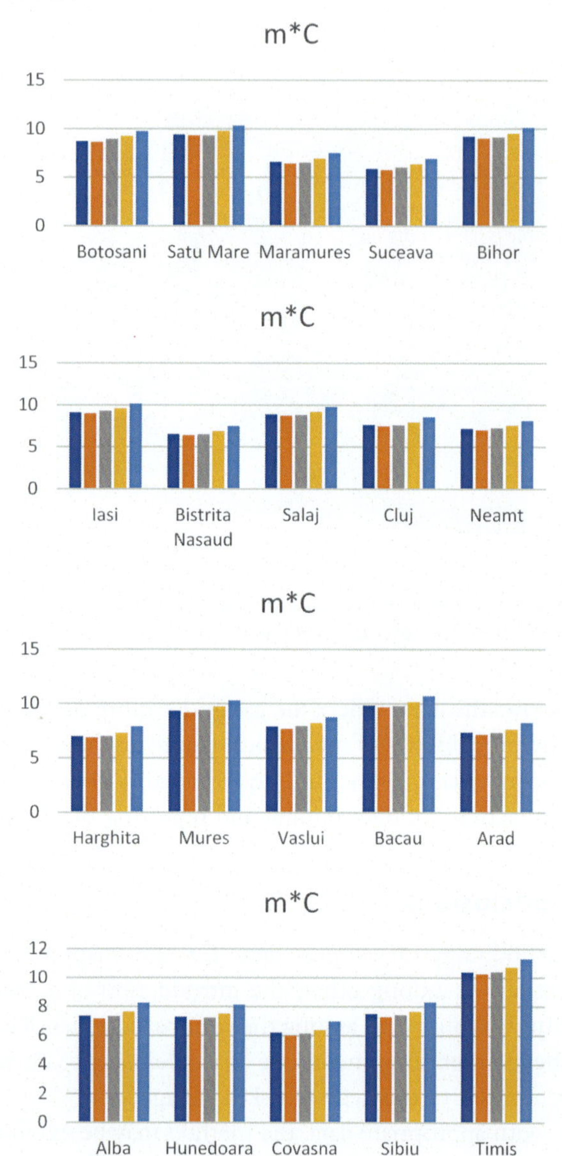

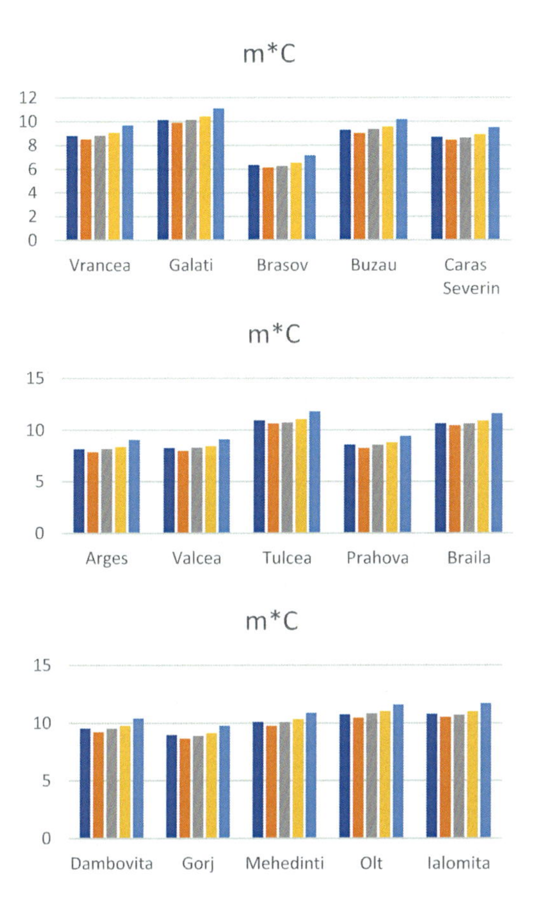
m*C
12
10
8
6
4
2
0
Vrancea
Galati
Brasov
Buzau
Caras Severin
m*C
15
10
5
0
Arges
Valcea
Tulcea
Prahova
Braila
m*C
15
10
5
0
Dambovita
Gorj
Mehedinti
Olt
Ialomita

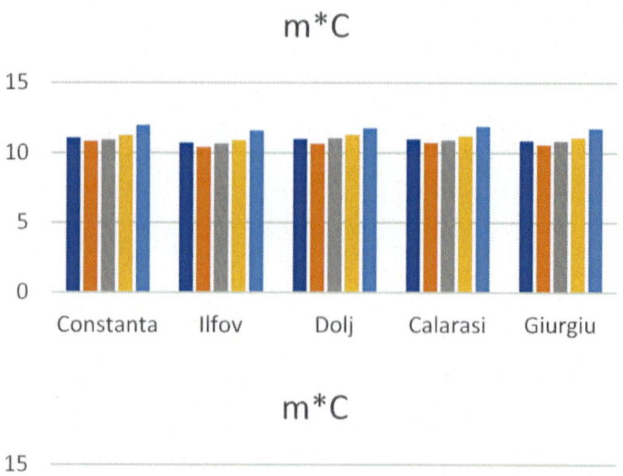
m*C
15
10
5
0
Constanta
Ilfov
Dolj
Calarasi
Giurgiu

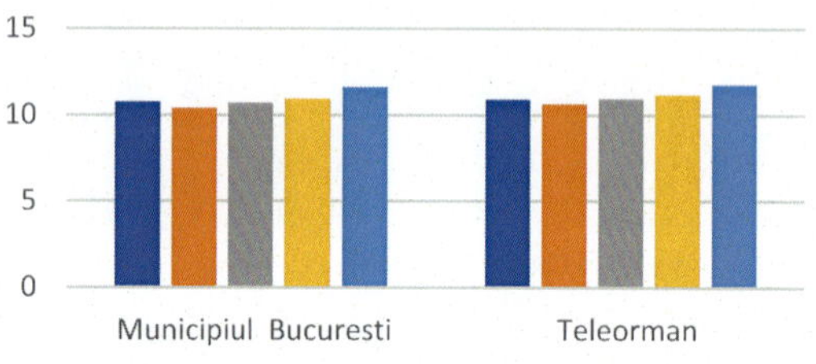
m*C
15
10
5
0
Municipiul Bucuresti
Teleorman

Standard deviation values.

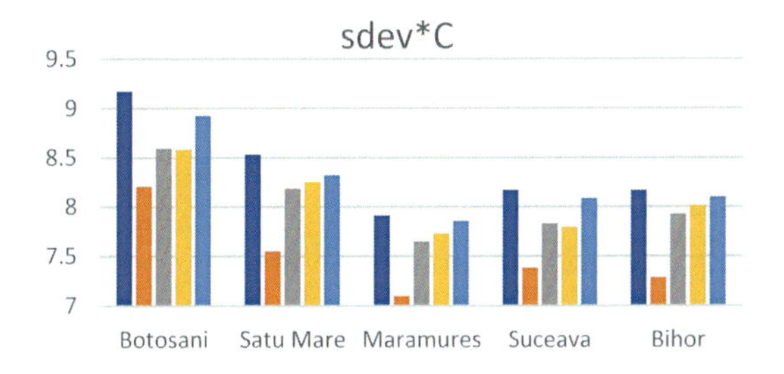

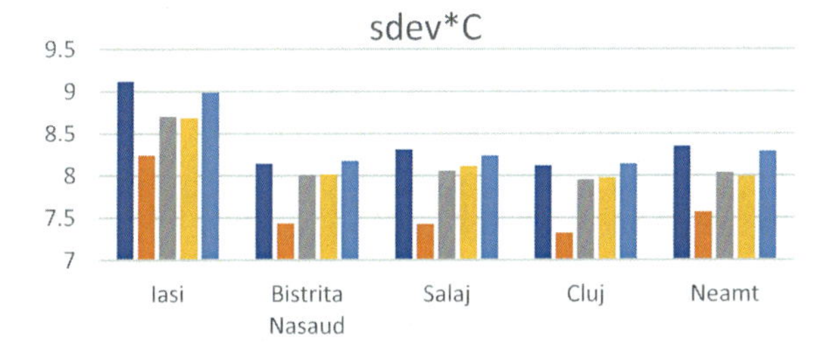

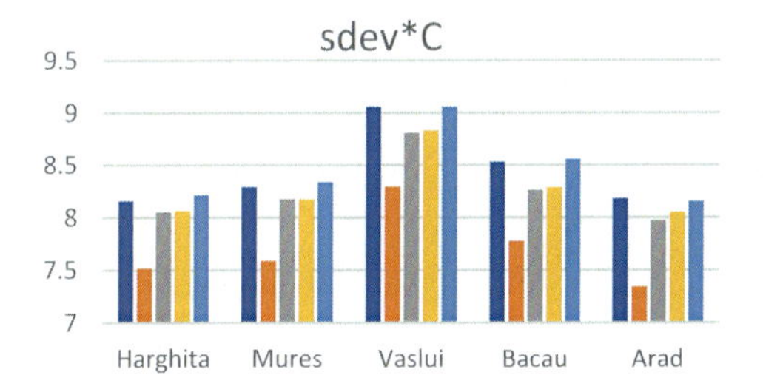

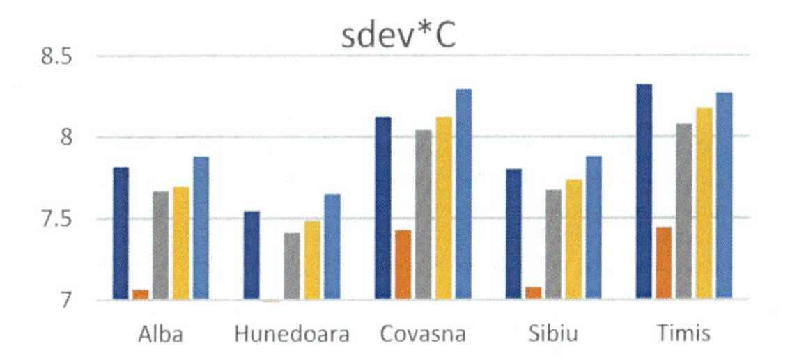

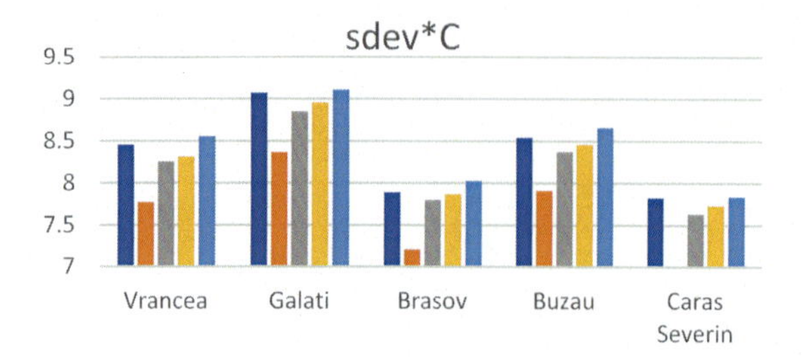

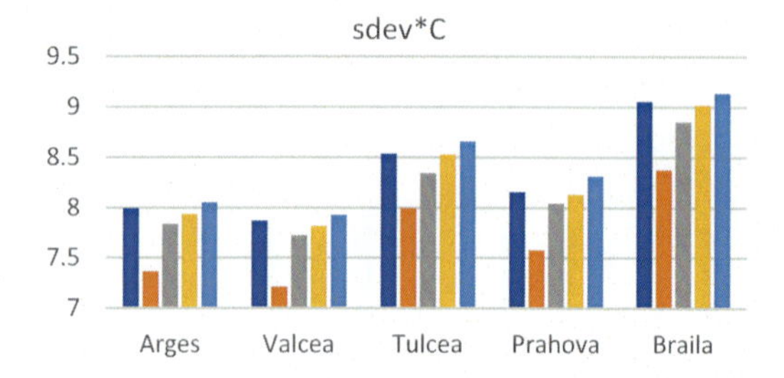

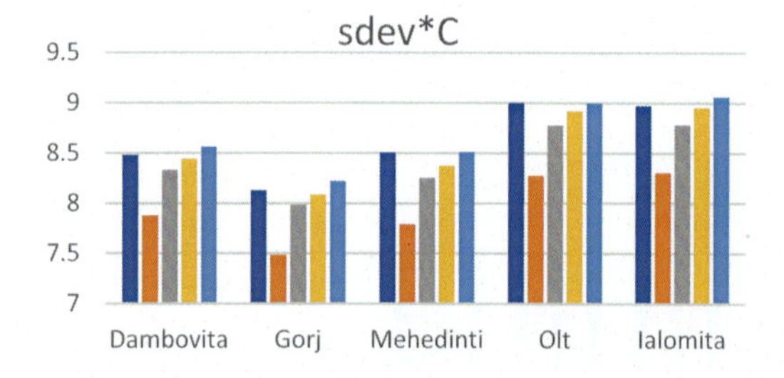

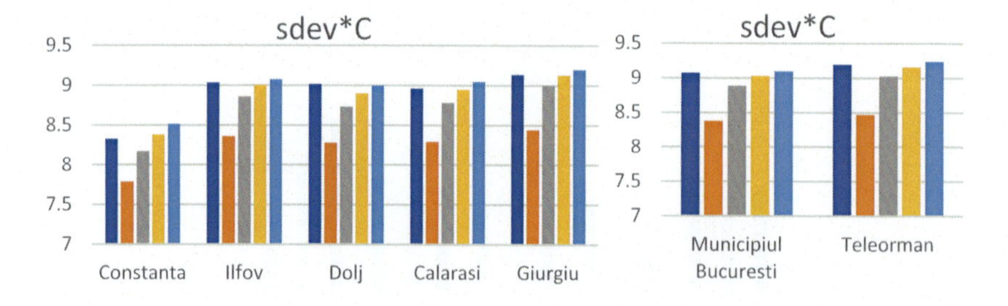

Annex 4.2

The analysis of the average temperatures shows that the values in the 1960s were larger than the ones in the following period. This is just a comparison of values without an assessment of the causes. This evolution shown

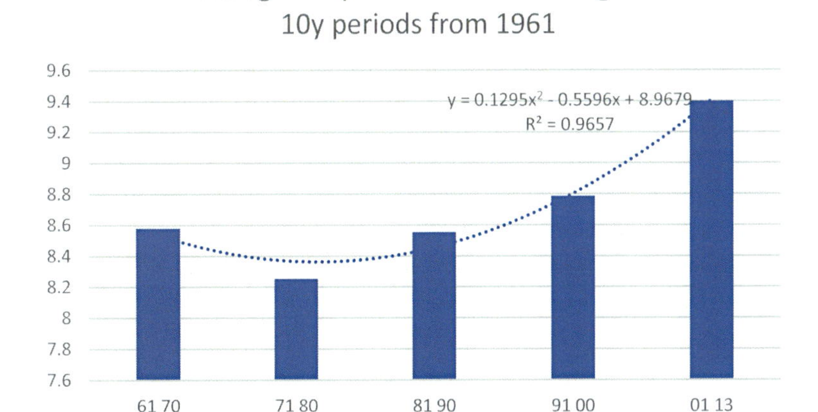

Figure 4.2.1 Prahova region average temperature evolution.

in Fig. 4.2.1 leads to the best fit of the evolution given by a second-order polynomial (see the formula in the figure).

Looking at the dispersion of the above distributions one may see the same trend of increasing values. In this case, the best fit is a fourth-order polynomial having an increasing part toward the future years (see Fig. 4.2.2.).

Future values of average and dispersion for the 2014–13 period may result from the regression functions determined above. In the case of the

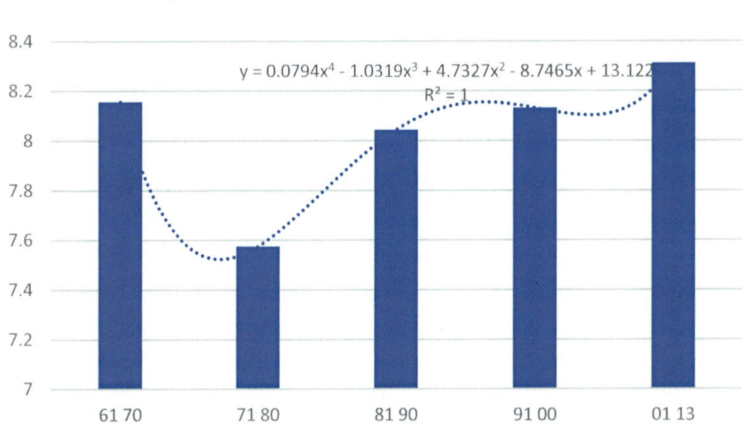

Figure 4.2.2 Temperature distribution dispersion values.

Prahova region, they are m = 10.2723 and s = 11.0322. These values will be useful in calculating the put option value that represents typical insurance for the risk of temperature volatility increase.

Put option as insurance.

Let us consider that there is a given value for the risk that is normally considered proportional to the dispersion of the temperature distribution. One may assume that the consequences' value is a component of risk along with the frequency of the event. In the case of insuring to a given risk value, the insured is interested in receiving compensation not lower than the value of the insurance. Using the derivative options language this represents a put option. This option is calculated using the formula given below:

$$S * (N1-1) - E\exp(-rt)\,(N2-1)$$

Where: S-strike price; E-exercise price; s-volatility; t-time; r-discount rate. And the values of N1 and N2 are the normal distribution (N) values for respectively N1 = N(0,1,d1) and N2 = N(0,1,d2) with d1 = (ln(S/E) + μ)/(s * sqrt(t)) and d2 = d1 − s * sqrt(t) and μ = r−0.5 * s^2 * t.

In the formula above, s represents the dispersion calculated above normalized by the average from above for the temperature distribution and t is the period, i.e., 10 years. Considering a typical value for r = 8% (0.08) and the value for s = 11.0322/10.2723 = 1.074 one has the values for μ = 0.08−0.5 * 1.074^2 * 10 = −5.687 and d1=(ln(S/E)−5.687)/(1.074 * sqrt(10))=(ln(S/E)−5.687)/3.39 = 0.29ln(S/E)−1.67;
d2 = d1−3.39 = 0.29ln(S/E)−5.06.

With these values, the put option price, in the case of an S/E = 0.5, would be

$$1 * (N(0,1,-1.87)-1) - 2\exp(-0.08 * 10) * (N(0,1,-5.26)-1) =$$
$$1 * (0.069-1) - 2 * 2.22 * (3.9e\text{-}7-1) = 3.509$$

For S/E = 0.8 the cost of the option is 0.086, i.e., 8.6% of the exposure. Considering the risk/cap of $196.16 from Table 4.2 for Prahova County, we have a cost of $16.9. We can do this exercise for all county results in Table 4.2 with the insurance premium.

References

EU Solidarity Fund. http://ec.europa.eu/regional_policy/thefunds/solidarity/index_en.cfm#1.

European Commission, 2007. Adapting to climate change in Europe - options for EU action. In: Green Paper from the Commission to the Council, the European Parliament, the European Economic and Social Committee and the Committee of the Regions. June 29, 2007.

Pollner, J., Kryspin-Watson, J., Niewwejaar, S., 2010. Disaster Risk Management and Climate Change Adaptation in Europe and Central Asia. The World Bank. http://www.preventionweb.net/files/15518_gfdrrdrmandccaeca1.pdf.

Purica, I., 1991. La dinamica dei Rischi nel Sistema Gas Naturale in Italia, Report in Program VESE. ENEA, Roma.

Purica, I., 2010. Risk dynamics in the Italian natural gas system. In: WEC Regional Energy Forum – Foren 2008, Neptun, 15-19 June 2008, Reference No: S4-29-En.

Purica, I., 2014. Report on Hydrological Basins Distribution of Population. ANAR, Romania.

Further readings

Ahmed, K., Sanchez-Triana, E. (Eds.), 2008. Strategic Environmental Assessment for Policies – an Instrument for Good Governance. The World Bank, Washington, D.C. ISBN: 9780821367629.

Hazard Risk Mitigation and Emergency Preparedness Project, (IBRD-47360 TF-53472), Report No: ICR23984.

IBRD, Hazard Risk Mitigation and Emergency Preparedness Project, (IBRD-47360 TF-53472), Report No: ICR23984.

IPCC, 2013. Managing the Risks of Extreme Events and Disasters to Advance Climate Change Adaptation: Special Report of the Intergovernmental Panel on Climate Change. Cambridge University Press. http://www.ipcc.ch/pdf/special- reports/srex/SREX_Full_Report.pdf.

ISTAT - Italy, and INS - Romania, (Sources of Data at National and Regional Level).

Maasten van Aalst, 2006. Managing Climate Risk. Integrating Adaptation into World Bank Group Operations. World Bank.

MARSH, 2005. Risk Assessment of Transelectrica Romania, Report.

Natural Hazards Assessment Network (NATHAN), a website that extracts information from Munich Re Group's NatCatSERV ICE database provides country profiles that include socioeconomic and hazard data. http://mrnathan.munichre.com/.

Palisade plc., @Risk, Decision Suite Program, Version 6.2, 2014.

Purica, I., 2015. Climate Change events induced risk assessment and mapping as a basis for an insurance policy. In: 2nd International Conference 'Economic Scientific Research - Theoretical, Empirical and Practical Approaches', ESPERA 2014, 13-14 November 2014, Bucharest, Romania, Procedia Economics and Finance, vol 22, pp. 495–501. https://doi.org/10.1016/S2212-5671(15)00245-2.

The European Commission Joint Research Centre. Climate Impacts in Europe. The JRC PESETA II project. http://ipts.jrc.ec.europa.eu/publications/pub.cfm?id=7181.

UN/ISDR - WB, 2008. South Eastern Europe Disaster Risk Mitigation and Adaptation Initiative: Risk Assessment in South Eastern Europe - A Desk Study Review. United Nations, Geneva, Switzerland.

Watts, A.C., Ambrosia, V.G., Hinkley, E.A., 2012. Unmanned aircraft systems in remote sensing and scientific research: classification and considerations of use. Remote Sensing 4 (6), 1671–1692.

World Bank, 2012. Turn down the heat. Why 4 degrees warmer world must be avoided. In: A Report for the World Bank by the Potsdam Institute for Climate Impact Research and Climate Analytics. November 2012.

World Bank. Insurance against Climate Change. Financial Disaster Risk Management and Insurance Options for Climate Change Adaptation in Bulgaria.

CHAPTER FIVE

Brief considerations of economic indicators

You can't use an old map to explore a new world.

Albert Einstein.

This fifth chapter describes indicators of climate change within the greater context introduced in the book, with no case studies but various examples of integrated indicators extending beyond the typical GDP/cap or tCO_2/cap metrics (i.e., two simple indicators). The chapter underlines the need for integrated indicators that reflect the evolution of the complex intercorrelation with the environment. The oscillatory behavior of GDP components is analyzed to support the contribution of economic innovation cycles to the shortening of Kondratiev cycles. Macrolevel circular economy indicators are summarized.

A brief presentation is done on variousVarious green investment schemes for implementing sustainable projects are briefly presented in the sixth chapter.

As a science, economics has passed through the epistemic stages of other sciences such as physics. When Tycho Brahe was noting the time series of data on the movement of the Moon, he was process observing Nature. Later theory was based on his data, and Newton and other physicists built theories that generated indicators from which the behavior of complex systems was analyzed and predicted.

Going with Purica (2010), we may say that the economy nowadays has a huge data set of time series on various dynamics and has developed strong methods to statistically analyze these data. Due to its intrinsic nonlinearity, only some linear models have emerged that could use linear indicators to predict economic behavior. Nonlinearities have been observed in economic systems and specific models have tried to cope with them. One example is the Kondratiev cycles that identified oscillatory behavior, another more recent example is the attempt to use complex dynamics for the description of discontinuous behavior (S. Keen, I. Purica, Benhabib, etc.).

As yet, no set of nonlinear indicators has been officially adopted for general use. The international institutions operating the databases mentioned

Climate Change and Circular Economics
ISBN: 978-0-443-29969-8
https://doi.org/10.1016/B978-0-443-29969-8.00002-1

have gathered several indicators of various aspects of economic and social activity.

In recent years anthropic activity has become a significant driver in the interaction with the environment and its impact is showing that some limits of the usual basin of behavior are close and assessing what may happen if they are crossed is not an easy job due to the lack of a complex view of the economic systems as one of the planets' interlocked systems.

5.1 From simple to aggregated

The simplest method for obtaining an image of the dynamics is to show the evolution of each indicator separately. Although this offers a good way to predict the future values of each indicator, it does not allow us to foresee the global system behavior.

To characterize the interaction, we have to start using combined/aggregated indicators. If, for example, in the hydro-dynamical systems we are very much accustomed to criteria numbers (Reynolds, Prandl, etc.) resulting in a combination of the system parameters (indicators), in the economic systems, although highly dynamic, the normally used characterization of the interactions is done by aggregated indicators resulted from the simple division of only pairs of the simple indicators (e.g., energy/capita, energy/GNP, CO_2 emissions/capita, CO_2 emissions/unit of energy, GNP/capita, etc.). We must note that the simple division indicators' evolution is easy to verify intuitively, thus, even if no model sustains the interpretation of the indicator related to data, one uses intuition to conclude the future behavior of the economic system.

In this context, the passage from a linear economy (describing the intercorrelation with Nature as obtaining resources from the environment—using them in the economic system that adds anthropic value—and then reintroducing the waste to Nature), to a nonlinear (circular) economy where new technological information is used to diminish and diversify the resources taken from the environment and where new anthropic value is added for products that also have a natural value-added embedded from the beginning of resource use such that the amount of waste (to humans) is either diminished or has a less destructive impact on the environment.

Thus, the various fluxes that keep the economy going such as money, energy, information, products, and labor have several types of indicators associated with them. For example, labor is described by indicators related to education and health, money by financial indicators, energy by generation and consumption, information by research and technological indicators, etc.

Linear economy models use only linear indicators, be they single values of economic parameters or ratios of such values or linear combinations of them.

If one wants to consider circular (nonlinear) economic models, the indicators must reflect the cyclic and more complex nonlinear behavior of the economy. Moreover, these indicators should reflect the intercorrelation between the various dynamic fluxes mentioned above. As an example, one may consider the correlation between economic production (described by a Cobb–Douglas formula) and the production of technological information through research, shown to have a similar description with different meanings of the variables (Purica, 2010), of which a basic diagram is given in Fig. 5.1.

The indicators of linear economies always have a positive trend to mark development and a negative one to show regress. Sometimes the wish for growth as the normal trend of the economy leads to describing the periods of downturns in evolution as "negative growth." This associates the idea of growth with that of progress, although the growth in waste thrown into the environment does not exactly mean progress.

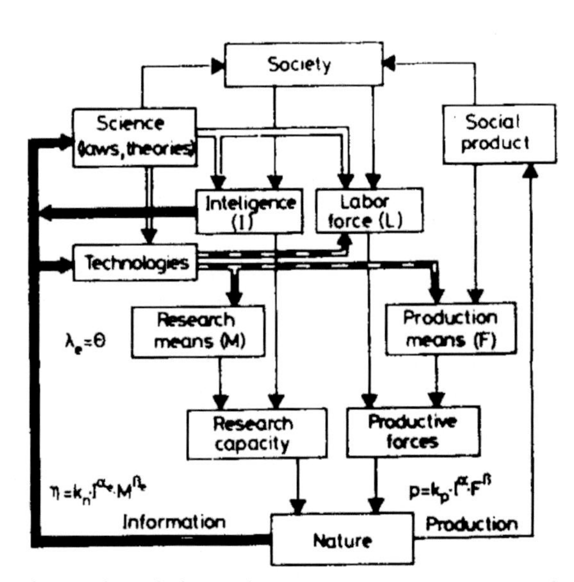

Figure 5.1 Production knowledge and GDP. *From Purica, I., 2010. Nonlinear Models for Economic Decision Processes. Imperial College Press – World Scientific, London, 2010, ISBN: 978-1-84816-427-7.*

So creating a more complex indicator (again, not related to a specific model) would require the use of more linear indicators in a formula that would include positive evolution indicators (e.g., GDP) at the nominator (a positive trend if they increase) and negative evolution indicators at the denominator (a positive trend if they decrease). This way, the complex indicator would increase if the evolution were positive (e.g., high GDP, low waste) and decrease in the opposite case.

Such an indicator may be, e.g., the product of energy productivity [GDP/TPES](1/energy intensity), GDP per capita [GDP/cap], and the rate of resource recycling [RR%] divided by the amount of CO_2 emissions [tCO_2].

This indicator would be measured in [$Euro^2$.%/kWh.cap.tCO_2]. One important comment here relates to the dimensions of indicators. For various indicators, little attention is given to dimensionality; thus, sometimes "apples are added to pears" without caring about the meaning of the resulting numbers. One way to resolve this issue would be to consider the sensitivity measure of the system (we are using the language of control systems theory now). Since one wants to describe the dynamic of economy-nature systems, sensitivity indicators may be more suitable, e.g., measuring the variation in reported GDP to the value of GDP (dGDP/GDP). In this case, the resulting number has no dimension, and the complex indicator would reflect this.

The passage from linear to circular views requires a change in mentality. The perception of change in complex systems, and accordingly the reaction, shows bifurcations-like behavior, especially when one acquires a conscience of the limits (environmental, technological, social, etc.). Linear mentalities like "whatever you do will change nothing" or "anything you do will change everything" must be changed to more subtle approaches that account for second-order effects that characterize the behavior of the complex systems we face (e.g., actions like dam-building to protect from rising sea levels should consider that the production of cement for the dams represents a source of CO_2, which contributes to a rising sea level.)

In addition, environmental-specific conditions need to be considered. The case of a highway built in the desert to progress from the old transport of goods and persons between two cities with caravans of camels is relevant when after 10 years, sandstorms cover the highway and transport continues to be done with camels. A similar example is a river that passes through an arid land zone with a traditional requirement to maintain a thick layer of trees along its borders. When, for various reasons, the people cut the trees below the traditional limit to a very thin layer, the river was drawn, and the desert extended over the existing land crops.

An intuitive (mechanical) representation is shown below.

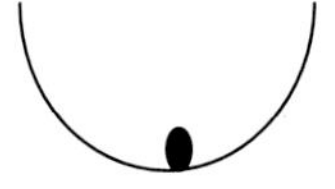

"Whatever you do will

change nothing"

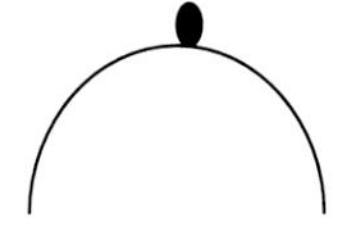

"Anything you do will change

everything"

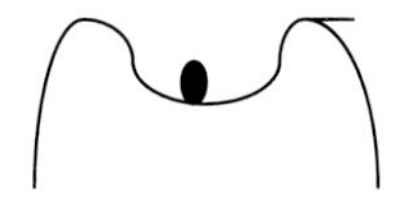

"Certain things you do will push the system

beyond stable equilibrium, others no"

During the Middle Ages, the indicator of welfare was the quantity of gold one possessed. Accordingly, the "research programs" of those days were aimed at changing everything to gold. After the occurrence of energy availability limitations, the indicators have changed. Additionally, the increase in complexity of interactions among various systems (energy, population, economy, environment, etc.) has led to the introduction of aggregated indicators. The planetary view we have today requires the consideration of the meteo-geographical conditions for each region and the normalization of the specific indicator values for making better comparisons. This suggests a personalization of the new technologies for energy supply being implemented in the various regions, which should consider not only the geographical conditions but also the social ones to achieve maximum efficiency.

The input of energy into an economy is increasing the production of GDP through the creation of infrastructures. Thus, the well-known bend in the energy intensity evolution may be represented as a typical investment process where energy is invested to create infrastructures that would increase the production of GDP per unit of energy. Correlating this process with the second-order effects in population migration and with the evolution of the CO_2 emissions per capita leads to straightforward conclusions. One of these is that the interconnection of energy systems (electric, gas, oil, etc.) is an important step in the creation of a global approach to development.

Making decisions for development has always been based on some type of representation of the process. Various models have served as tools to devise or justify decisions. The mathematics behind those models was, usually, linear. Since the behavior of the processes involved was highly nonlinear, the approximations made were valid for restricted areas and restricted time intervals. These models were not able to predict the limits beyond which a discontinuous behavior should occur in systems evolution. Decisions of the type "quit financing a technology and enhance others" are common in the economy. Only in the last years nonlinear models, based on nonlinear mathematical tools, have made possible the prediction of discontinuous decisions occurring when certain system parameters cross some limit. Although the mathematics involved is more complicated than the linear one, the representation of systems evolution among limits is more straightforward.

Several approaches have been tried based on either biophysics system modeling applied to economies in correlation with nature or physical systems' models such as gauge fields or quantum mechanics mostly applied to financial systems. Nonlinear models that can generate deterministic chaos regimes have also been devised along with reaction-diffusion systems descriptions.

One interesting class of metrics stems from irreversible thermodynamics modeling of the exchange of exergy between economies and nature along with the material resources. This approach considers the generation of entropy that results from the thermodynamic equations and uses it as a relevant indicator for the interaction nature-economy. One result of such models points out that there are limits that have to be considered associated, e.g., with each resource and technology (the useful energy—exergy—is one such limit), as well as to the entire planet. Saturation toward reaching these limits makes the logistic curve description more adequate for the planetary scale of our economic systems' behavior.

Even if not accepted, limits may, sometimes, be avoided or, seldom, crossed, with the associated shocks. The capability to absorb shocks and still perform normally (resilience) measures the impact of our decisions for development on the environment, the economy, etc. Accepting the limits, as another alternative, opens the way to understanding the mutual interactions among the various systems, becoming, thus, able to change those limits in a sustainable symbiotic evolution.

Negotiating economy versus environment in our development involves information that is not always available and time constants that may be longer than what we have presently dealt with. The costs and financial measures, implied, may lead, for example, to capital accumulations that we are not prepared to control yet, requiring appropriate administrative structures. The present changes in, e.g., energy generation, transmission, distribution, and use systems lead to more players in the market who add to structural problems related to the role of a regulator that would avoid the occurrence of chaos in the process and thus avoid shocks to the economy. Correlating global climate change with energy is one of the first projects that considers the interactions among various systems at a planetary scale, opening the way for closer international cooperation.

5.2 Normalization of parameter values

The fact that any country is characterized by its average temperature (t_m) influences the specific energy consumption for heating and other temperature-dependent activities. Temperature dependence is inversely proportional (i.e., the higher the temperature, the lower the energy consumption for heating).

Similarly, we may consider that the higher the population density (ls) is, the lower the energy consumption for transportation will be.

Comparing the energy consumption of different economies, one must eliminate the influence of meteorological-geographical factors to have a realistic view of energy consumption to sustain an economic structure.

The carbon-equivalent emissions of a country with a larger forest surface will produce a different environmental impact than a country with little forest surface due to the absorption capacity of the forests. A comparison of emissions that contribute to producing environmental impacts must account for the forest absorption effect.

The above comments suggest that normalization for some reference values should be completed to obtain meteorological-geographical free comparable values for energy consumption and carbon-equivalent emissions.

5.3 Population migration—A potential cyclic behavior due to saturation

The problem under study nowadays is the migration of population (in and out of various areas) relative to the economic characteristics (like infrastructure and GNP) of respective areas.

Now, what is the potential gap that produces a variation in the concentration (a flux) of persons from one area to another? With the political transitions we are witnessing, we shall consider that people are simply going from poor to rich areas. So the difference that is intensively perceived between West and East or between North and South is one of welfare/poverty.

Considering the welfare/poverty barrier between West and East or North and South and the consequent in-migration of the population from poor to rich areas, we may identify nonlinear (cyclic) behavior like the one described below.

The infrastructure measures the efficiency with which an economy makes labor (active population) produce GDP, being expressed as GDP/capita.

Increasing the population, by in-migration, represents an increase in the active population (labor). Over a certain saturation value of the infrastructure's efficiency, the increase in population shall be greater than its capacity to produce GNP. So, the GNP per capita will diminish, this being perceived as poverty. Thus, along with in-migration from the poor to the rich areas, there is also a long-term import of poverty into the rich economy.

In parallel, the investments of rich economies for creating (or developing) infrastructures in either the east or the south contribute to increased efficiency in those economies. Thus, the increase in efficiency will produce a greater GDP/capita, perceived as an import of welfare from the rich economy to the poor one.

If this perception is strong enough the outflow of population might reverse. These reversions may create cycles of in-out migrations from initially poor regions where investments are made in developing the infrastructure. A good enough example is provided by Italy where the out-migration of the 1950s reversed in the mid-1970s (see Fig. 5.2.), this being a sign that the infrastructures were set and operational.

The same trend is seen later with Turkish migration to the EU and hopefully will be seen in the future with present-day migrants.

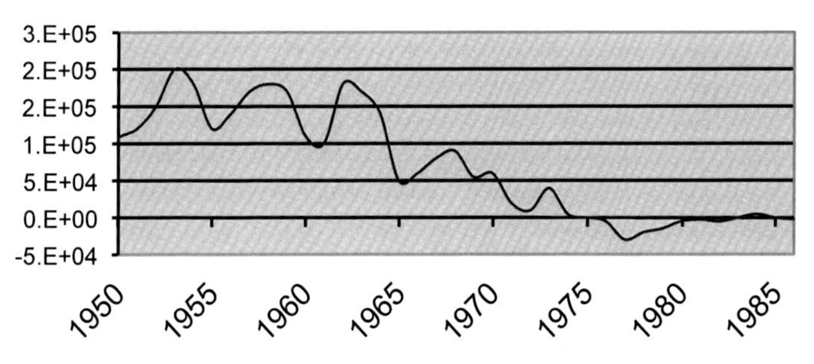

Figure 5.2 Italy EX-In patriated. *From ISTAT.*

The existence of limits that may generate cycles in the economy-nature interactions raises the problem of coping with these limits and reflecting them in circular economy indicators.

5.4 Avoiding or crossing limits—System resilience

When limits may be predicted one would be tempted, in the first place, to avoid them. On second thought, after having estimated the shock to the economy, crossing certain limits may become a better decision. Thus, if the systems capability to absorb shocks (resilience) is good enough, then, the decision will have to specify not only where to invest but also when so that certain limits or preselected conditions are avoided or not crossed.

The possibility of the nonlinear approaches to include the moment as an element of the decision is more similar to what we are faced with in day-to-day life, giving a higher predictability level for these models.

5.5 Sustainability—Accepting limits

Once the consciousness of the system's dynamics is created in a systematic way to show how to control the evolutionary trajectory, another possibility occurs, representing an interaction of a superior level. It refers to the fact that by designing the parameters of a system one may control the position and amplitude of the limits. Thus, the alternative of accepting the limits and trying to control them by influencing the parameters proves to be the best long-run decision. The one word that describes this mentality is "sustainability." This approach might lead to one to better cope with the reality of changes in climate and energy conversion patterns.

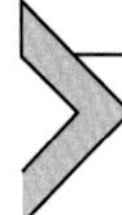

5.6 Economy versus environment—Negotiating development

Assuming that governments decide (after the Paris COP 21 agreement) to respond to the reality of climate change, there is some uncertainty about how they will respond, e.g., to follow the example of the EU's green deal to reach emissions neutrality. Especially two basic types of response are possible. One is to attempt to prevent the impacts ex ante (i.e., the anticipatory prevention approach—mitigation). The other is to attempt to adapt ex post to the changes in climate in the most efficient manner possible (adaptation approach). Obviously, both responses are expected.

5.6.1 Information

Under the anticipatory prevention scenario, various actions would be taken to reduce CO_2 and CFC emissions to confine future climate warming to within an "acceptable" level. If, for example, the fuel mix is changed compared with what it would have been in the absence of climate change, we may presume that the change has positive costs (otherwise it would have been chosen anyway).

Under the adaptation scenario, various policy options exist. Adaptation may be as obvious as building higher sea-wall defences, or as broad as making changes in land use practices. Expenditure on this type of policy will ameliorate the impacts of climate change after the changes have occurred.

The uncertainties about climate change are both scientific and sociobehavioral. We do not know enough at the moment to say what the average global temperature and the sea level rise will be. We know even less about the dynamics of spatial distribution of the temperature rise, and hence an analysis of the regional impacts is necessarily speculative. Moreover, impacts depend on how people respond to climate warming and sea level rise and on the kind of actions that governments will take. All this suggests, strongly, that there has to be a great deal more of both scientific and socioeconomic research on climate change. One example is the amplifying cycles like protection from the rising sea level, induced by emissions increase, with more cement dams that need more cement whose production increases emissions thus inducing more sea level rise.

5.6.2 Time

But now consider the policy implications of this uncertainty. Does it, for example, mean that we should delay action until scientific and socioeconomic research narrows the range of uncertainty? There are reasons for acting now: first, the outcome, if the worst happens, is clearly catastrophic. Unless we are positively in favor of bearing risks, we need to undertake risk-averse policies. Second, the longer the world community delays action on the greenhouse effect the greater will be the "committed" level of warming. Delay is therefore not cost less: the damage from greenhouse gases will simply be greater the longer we delay. Third the design of efficiently targeted policies should certainly wait for further information—it would be economic folly to engage in costly policies now if we can target them more efficiently later on. But a rational approach is to begin with low-cost policies to contain greenhouse effects—energy conservation is a conspicuous example. Fourth, prevention must be international. However, procuring international cooperation has time lags. Indeed, the speed with which the Montreal Protocol on CFCs was obtained may be the exception, not the rule. The costs to CFCs are insignificant compared with what must happen if the greenhouse effect is to be truly confined. There are reasons to believe that, even though the global benefits are large, the existence of substantial control costs could make global cooperation very difficult: countries may not know if they will be gainers or losers, and some countries will partly gain and partly lose, and that we cannot say when these gains and losses will occur, and the difficulties of securing international agreement build even further. The past years' turmoil in the continuation of the Kyoto Protocol (finally accomplished by the Paris Agreement) is an unwanted example of the above.

5.6.3 Indicators of sustainability

At this point "sustainable development" comes in as a fundamentally different approach shifting the focus from economic growth as narrowly constructed in traditional attitudes to economic policy. It speaks of development rather than growth, of the quality of life rather than real incomes alone. That is, sustainable development makes it clear that the very antithesis of growth and the environment is not the issue.

Sustainable development accepts that what we have been calling "economic growth" in the past has been measured by some very misleading indicators. The tendency has been to use a measure of gross domestic product (GDP) as the basis for economic growth calculations. If GDP increases that is economic growth. But GDP is constructed in a way that tends to divorce it from one of its underlying purposes: to indicate, broadly at least, the living standard of the population. If pollution damages health, and health care expenditures rise, that is an increase in GDP—a rise in the "standard of living"—not a decrease. If we use up natural resources then that is capital depreciation, just as if we have machines, we count their depreciation as a cost to the nation. Yet depreciation on man-made capital is a cost while depreciation of environmental capital is not recorded at all. Broad classifications of current indicators potentially relevant to the circular economy are summarized in Tables 5.1 and 5.2.

There is, definitely, a cost of environmental policy in various countries but, this is fortunately not the only thing that influences the public perception of "growth versus the environment" issue (Pearce et al., 1990).

The message in the above comments is that, from now on, along with the man-made, material, and "know-how," capital we have to seriously take into account the environmental capital. Finding and implementing indicators to show the depreciation of the later capital will not be an easy task. Implementing environmental standards will certainly bear important costs for all nations involved. Hopefully, the new emissions trading mechanisms are bringing in a commercial component that is very likely to alleviate those costs.

5.6.4 Costs

A necessary condition for reaching, at minimum cost, the environmental standards, established by many countries, is that harmful environmental emissions are distributed so that all sources with positive emissions of the material under consideration should have equal marginal costs of emission reductions. A uniform percentage reduction in emissions from all sources will normally lead to an emission pattern that makes the marginal costs of emissions differ between sources. This is true whether "sources" are interpreted as different firms or consumers within a country (i.e., national efficiency), or as different countries (i.e., international efficiency).

The conclusion tends to suggest that energy conservation measures are likely to be more cost-effective than technology substitution for initial CO_2 reduction measures.

5.7 Conclusions

Having considered the perception of the circular economic parameters as a complex process where certain actions that push the system beyond certain limits may lead to a change in its equilibrium, we have shown that the indicators are based on the economic process characteristics. The interaction economy-environment is described by extending the notions of capital production/destruction from only the man-made one to include the environmental one. This is a result of having acquired the consciousness of starting to reach some saturation limits in an environment used to consider both an infinite source and an infinite sink.

Within such a complex dynamic the use of simple indicators is not enough to allow one's perception of the process evolution. The static description is not sufficient.

As an example, the reversing of population migration is another proof of the perception, by the active population, of the increased efficiency in the production of GDP. This may be considered a second-order approximation effect used as a dynamic indicator. Since the stability of the economic evolution is reflected by the first derivatives of the indicators, and not by their values, the sign change of a flux (e.g., of population) may be a good dynamic indicator.

Further on, a normalization of some indicators is suggested with respect to the meteo-geographical parameters of each economy under consideration. If one is supposed to improve the comparability of the indicators then, such factors as average temperature, population density, etc. have to be eliminated to have a unitary view of the structures and infrastructures of different economies.

Negotiating our development nowadays is accomplished in a climate of high uncertainty involving policies with various timescales. The limits of our usual indicators confronted with the description of "sustainable development" announce a change in our way of interpreting the dynamic socioeconomic and environmental system of which we are a part. For a better view of development, the implementation costs of new emissions-reducing technologies are also presented.

To conclude, we note that socioeconomic and environmental systems, with their high dynamicity, offer a wide field for improving our understanding of nature and how we live with it.

In the case of cyclic economies, some indicators may show development when their values are decreasing and regression when those values are increasing. For example, a decrease in the quantity of resources taken from the environment means progress, while an increase in waste or GHG emissions shows regression.

In this context, the report on the circular economy should be a synthetic survey of primary linear economy indicators (e.g., population, GDP, energy, and GHG emissions) to provide the basis for an extended analysis of indicators describing processes specific to a circular (nonlinear) economy, such as the RR resource recovery.

This class of indicators may include basic economic parameters at various powers and result in a different ranking of various economies from their present ones based on linear indicators.

An example is a combination of economic parameters such as GDP/En * GDP/pop. * RR/GHG = GDP^2/(En.pop.GHG) measured in USD^2/(kWh.cap.tCO_2).

One may even identify domains of constant behavior and limits of critical change (such as the 2°C).

5.7.1 Measuring circular economy—complex indicators

Table 5.1 Macro-level circular economy indicators (Geng, 2012).

Category	Indicators used
1. Resource output rate	Output of main mineral resource, output of energy.
2. Resource consumption rate	Energy consumption per unit of GDP, energy consumption per added industrial value, energy consumption per unit of product in key industrial sectors, water withdrawal per unit of GDP, water withdrawal per added industrial value, water consumption per unit product in key industrial sectors, coefficient of irrigation water utilization.
3. Integrated resource utilization rate	Recycling rate of industrial solid waste, industrial water reuse ratio, recycling rate of reclaimed municipal wastewater, safe treatment rate of domestic solid wastes, recycling rate of iron scrap, recycling rate of nonferrous metal, recycling rate of waste paper, recycling rate of plastic, recycling rate of rubber.
4. Waste disposal and pollutant emissions	Total amount of industrial solid waste for final disposal, the total amount of industrial wastewater discharge, the total amount of sulfur dioxide emissions, total amount of COD discharge.

Table 5.2 Broad classifications of current indicators potentially relevant to the circular economy.

Indicator type	Examples	Availability of data	Relevance to the CE
Sustainable Development	Social economic development, sustainable Consumption and production, social inclusion, demographic changes, public health, climate change and energy, sustainable transport, natural resources, global partnership, and good governance (Table A2)	Voluntary-based reporting via EU Directorate-general for energy (focused), European Sustainable Development Network (ESDN); corporate sustainability indicators (e.g., carbon disclosure)	Natural resources Sustainable consumption and production
Environmental	Agriculture, air pollution, biodiversity, climate change, energy, fisheries, land and soils, transport, waste, water	Regulatory-based reporting via EEA core indicators and country-specific statistics	Waste generated, packaging waste generation, and recycling
Material flow	Domestic extraction, direct material consumption, domestic material input, physical trade balance, net additions to stock, domestic processed output, total material requirement, total domestic output	Eurostat, SERI	All
Societal behavior	Sharing, municipal waste recycling, waste generated per capita (total and segregated), environmental/resource taxation	National and voluntary organization statistics	All
Organizationalbehavior	Material flow accounting in organizations, remanufacturing, use of recycled raw materials, eco-innovation, per capita statistics (e.g., reduction in waste generation per capita)	Private sector voluntary reporting via EU Forum for Manufacturing; ZVEI (German Electrical Industrial Association); VDMA (German Engineering Federation); etc.	All
Economicperformance	Resource productivity, recycling industry, green jobs, waste generation/GDP, "transformation of the economy"	EurostatEU resource efficiency scoreboard	All

Box 5.1 An illustration of a potential composite indicator

One possible combination would be as follows.

- energy productivity (GDP/total primary energy supply) where larger values are associated with progress;
- GDP per capita (GDP/population): the present indicator for progress;
- the rate of resource recycling (recycle rate as a percentage): improved recycling would increase this indicator.
- divide by the amount of carbon dioxide emissions, so reducing emissions would increase the indicator.

According to the formula

$$\frac{\text{GDP} \times \text{GDP} \times \text{recycle rate}}{\text{TPES} \times \text{population} \times \text{CO}_2 \text{ emission}}$$

Using data from IEA (2013) (except for the recycling rate), we obtain the results shown in the following table.

Country	Population (M)	GDP (trillion US$)	TPES (toe per capita)[a]	CO_2 emissions (Mt CO_2)	Recycle rate (%)	Composite indicator value[b]
USA	314.3	14.232	6.8	5074	37	2.2
Germany	81.9	2.851	3.8	755	45	19.01

Purica et al. (2016).

In essence, the circular economy concept should be accounted for to comprehensively relate the economy to the environment. We clearly must consider more than one indicator, i.e., emissions, to properly manage our interactions with the environment.

Circular economy strategies should be varied and encompass all domains and potential solutions.

A visual suggestion is provided below as one potential basis to underscore the complexity we face.

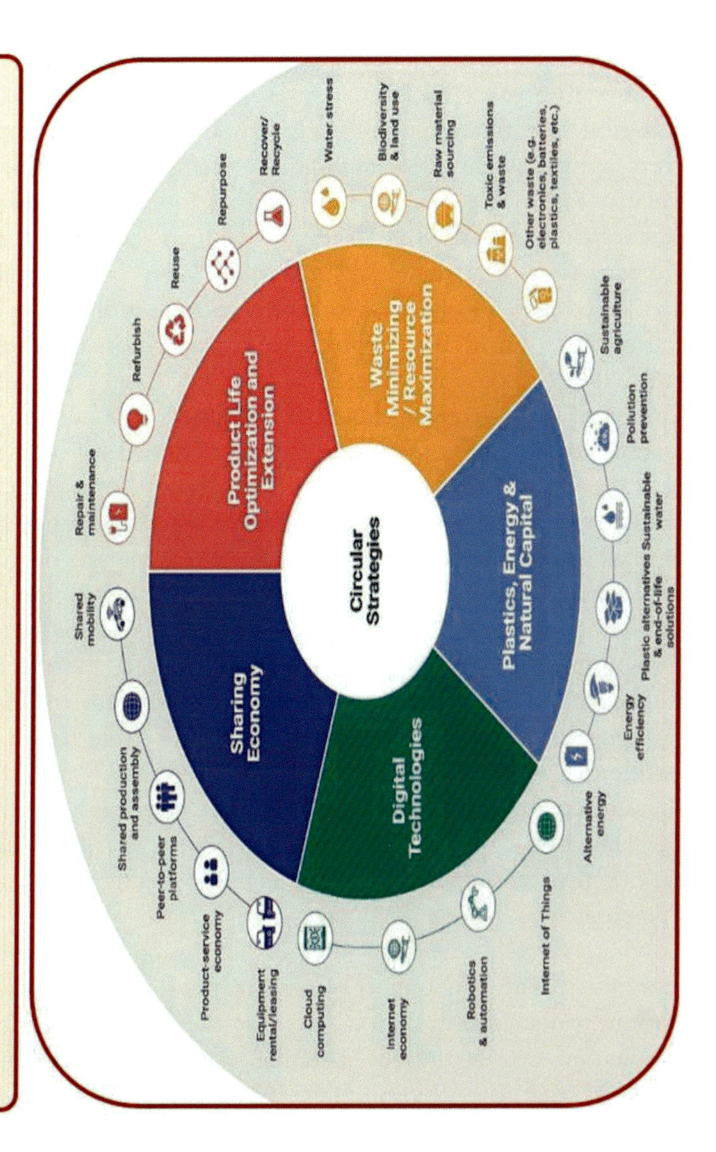
CIRCULAR ECONOMY STRATEGIES
Circular Strategies
Sharing Economy
Product Life Optimization and Extension
Waste Minimizing / Resource Maximization
Plastics, Energy & Natural Capital
Digital Technologies
Shared mobility
Repair & maintenance
Refurbish
Reuse
Repurpose
Recover/ Recycle
Water stress
Biodiversity & land use
Raw material sourcing
Toxic emissions & waste
Other waste (e.g. electronics, batteries, plastics, textiles, etc.)
Sustainable agriculture
Pollution prevention
Sustainable water
Plastic alternatives & end-of-life solutions
Energy efficiency
Alternative energy
Internet of Things
Robotics & automation
Internet economy
Cloud computing
Equipment rental/leasing
Product-service economy
Peer-to-peer platforms
Shared production and assembly

Several considerations should be underlined here.

1. A coherent organizational structure that brings the benefits of technological and financial knowledge to each project completed in every applicable country undergoes a change from a pyramidal to a matrix form. The columns of the matrix for each country or group of countries indicate the necessary domain expertise for specific projects. The knowledge resources are thus grouped into hubs that are available as needed to create optimal project development teams. The World Bank completed such a reorganization in the mid-1990s.
2. Commercial mechanisms should be designed with caution. A good example is the emissions trading scheme. The role of such a mechanism was initially to foster an overall reduction in emissions by introducing a commercial instrument called emission allowance units, i.e., tradeable certificates between various industries seeking the final result of reducing emissions by polluting entities to avoid paying for certificates, thus increasing investments into new less polluting technologies. When a zero emissions situation is attained, the mechanism is no longer needed. This is just theory; in practice, the trading scheme has generated large profits for companies that have traded emission certificates, which makes them reluctant to ever stop the mechanism. So, assuming that zero emissions will be reached, it is likely that a new mechanism will somehow remain that distorts the initial benefits to the economy and society. Commercial creativity is sometimes dangerous.
3. Kondratiev cycles describe the long-term constant evolution of economies. The penetration of new technologies in finance, information, production, transport, etc., has the side effect of shortening Kondratiev cycles, thus accelerating economic evolution. If one considers that the environment has a specific time constant of

recovery after each economic cycle during which resources are used and waste is reintroduced into the environment, faster economic cycles could be detrimental to economy—environment dynamics. An example of an economy's oscillatory behavior is provided in Annex 5.1. for Romania. Cesare Marchetti has done extensive research on innovation cycles during his activity at IIASA Laxenburg. One of his analyses of energy innovation is shown in the following graph. One may see that the next cycle that starts these years includes the prediction of fusion energy generation. We are also presenting a more colorful graph, done presently, of the innovation cycles just to stress the prediction capability of big data analysis done with a deep understanding of system behavior.

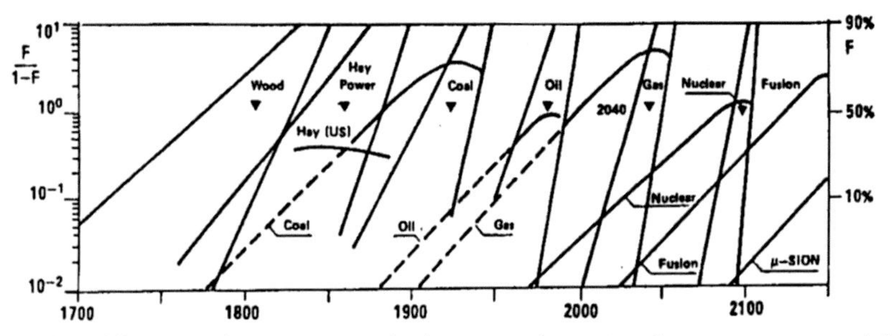

Reproduced from Marchetti, C., 1994, The long-term dynamics of energy systems and the role of innovation. International Institute for Applied Systems Analysis, Laxenburg, Austria.

The prediction extends to 2100, and assuming that each wave has a primer innovation technology, the last waves predict fusion and elementary particle energy generation as suggested by Marchetti. The basic model for this analysis is logistic and was used successfully for several domains' evolution predictions. I have used a similar model to predict nonlinear financial behavior, usually perceived as financial crises (Purica, 2015).

Innovation cycles are becoming shorter, as presented in the next table.

First wave	Second wave	Third wave	Fourth wave	Fifth wave	Sixth wave
Water power Textiles Iron	Steam Rail Steel	Electricity Chemicals Internal combustion engine	Petrochemicals Electronics Aviation	Digital network Software New media	Digitization (AI, IoT, AV, Robots & drones) Clean tech
60 years	55 years	50 years	40 years	30 years	25 years

The table is a synthesis of the following figure.

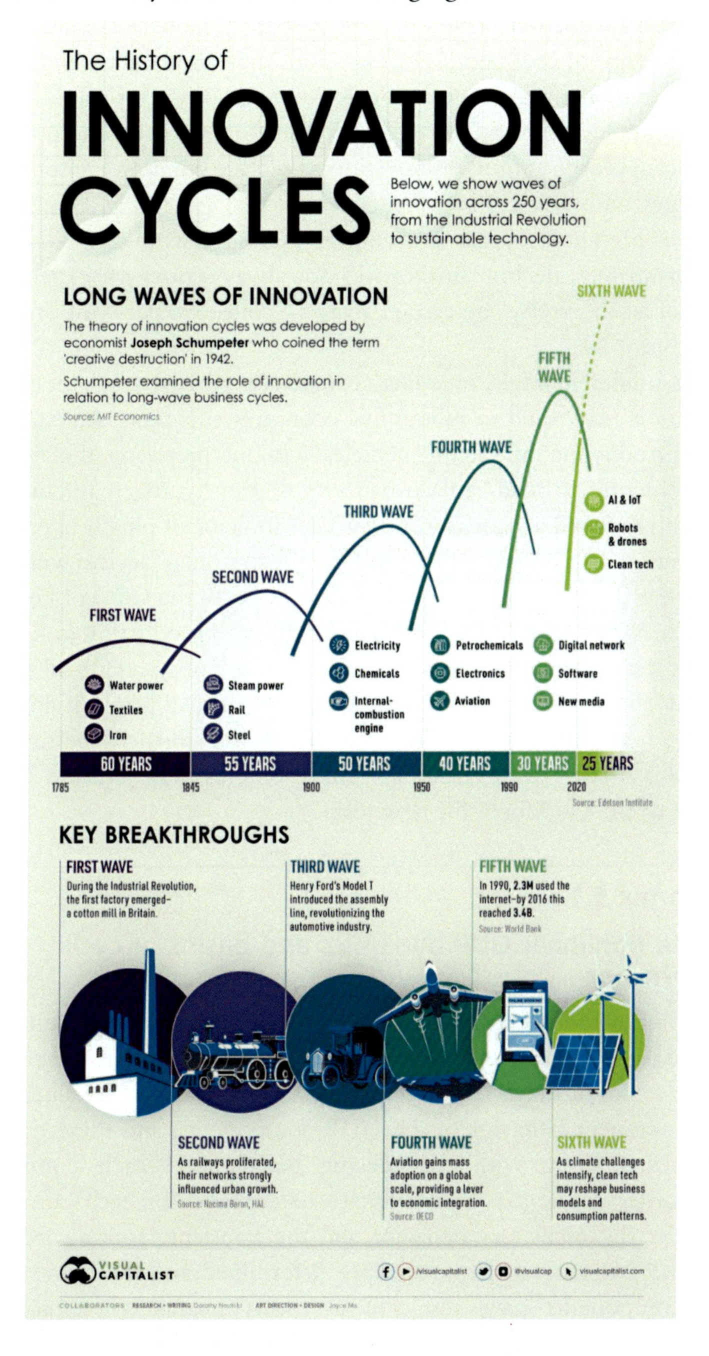

4. Considering only one indicator, in this case, emissions, may lead to resource use decisions based on flawed logic. The recent trend of stopping the use of coal because of emissions may also lead to stopping the use of transport because transport is the most important source of local emissions. How would an economy function without transport? The correct approach is to use innovation to replace emitting transport technologies with electric or hydrogen-powered vehicles and similarly to replace present-day coal-emitting technologies with clean-coal ones. Unfortunately, decision-makers do not always count on expertise but instead let themselves be carried away by collective behavior or specific interests.
5. From a different angle, speeding changes along based solely on emission indicators may result in suboptimal economic situations. An example is the introduction of electric vehicles without preparing the necessary infrastructure to load/replace batteries or the hydrogen for fuel cells, which produced a saturation in EV sales in favor of plugin hybrid EVs. Innovations for better batteries and hydrogen storage devices will reverse the trend. Some may recall an old interview with the CEO of Ford Motor Co. When asked why Ford had not changed production to EVs with plastic cases, he pointed out that conversion from present oil product supply chains and the metallurgical industry of steel plates, if done suddenly, may induce not only supply chain and financial perturbations but also social effects resulting from unemployment and the learning curve of the workforce for new jobs.

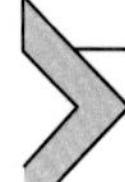

Annex 5.1

The nonlinear GDP dynamics and basins of cyclic behavior

Going with Purica (2012a) we may say that the oscillatory behavior of GDP and its components leads to a Fourier transform analysis that results in the eigenvalues of the dynamic economic system. The larger values dominate the transient behavior of the GDP components, and these transients are discussed along with the specific behavior of each component. Second-order differential equations are determined for each component to describe the oscillatory behavior and the transient resulting from step excitation. The natural frequencies are determined and the correlation of pairs of components' cycles results in the "beats" process where modulated wave cycles are compared and discussed. Nonlinear processes are reviewed, and the possible occurrence of dynamical behavior basins for GDP is explored. Important conclusions are drawn from this analysis of the dynamics of GDP and its components.

Introduction

Oscillatory processes are embedded in economic systems' behavior due to the intrinsic nature of human activities and the natural phenomena with which they interact. Various names have been coined for this behavior, such as cyclic, seasonal, yearly, quarterly, periodic, etc. Moreover, the scale of time constants of various activities ranges from seconds, e.g., the ticks of the stock exchange, to tens of years, e.g., T bonds of the US Treasury.

As shown in Appendix 1, various papers have identified nonlinear behavior and described it with various models that show features such as bifurcation, discontinuity, periodicity, etc. What we intend to present here is a systematic analysis of the oscillatory behavior of the GDP and its components that applies well-known mathematical instruments, e.g., Fourier transforms, differential equations, flows, etc.

By applying this type of analysis to the real GDP data from a given economy (Romania) we obtain a system of second-order differential equations. Each of these equations has its specific solutions. The resulting amplitudes of the Fourier analysis could be considered eigenvalues of these equations and the flow of trajectories may be discussed in terms of convergence/divergence, stability, and the underlying potential.

We have done a basic analysis of this sort (Purica, 2012b) in a previous paper on industrial production. In what follows we are extending the analysis to the GDP and all its components.

Data and Fourier analysis

The data of the GDP and its components are given in Appendix 2 as a quarterly series starting with Q1 2000. The source is the National Institute of Statistics of Romania. The values are retrievable from the site of the institute. We have taken the data given in current values, and performed the Fourier analysis on the series, taking the last 32 values of each, as required by the Fourier calculation.

The resulting amplitudes are presented in Appendix 3, while the graphs of the amplitudes are given below:

Industrial Production.

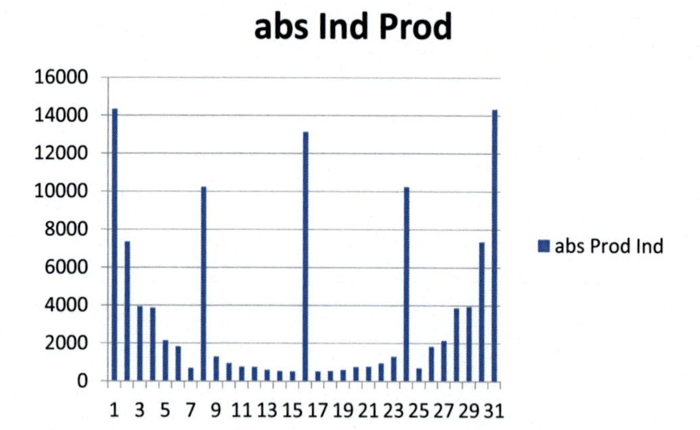

Commerce.

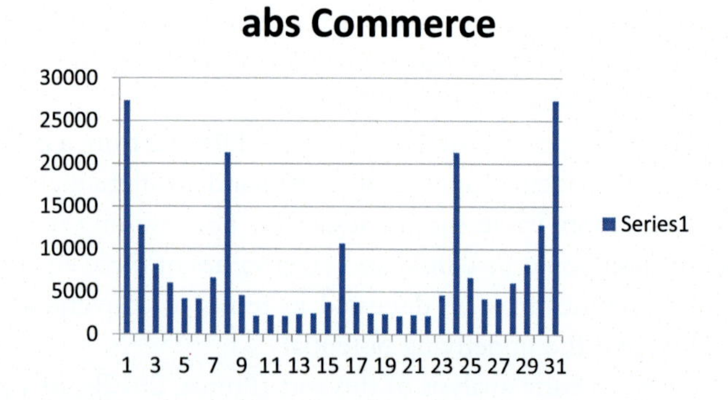

Construction.

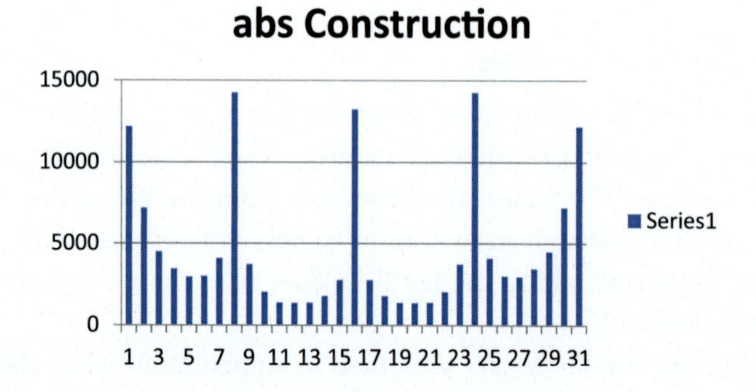

Finance.

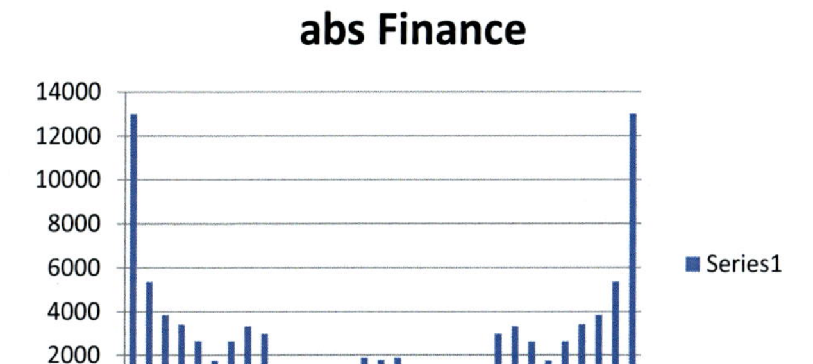

Services.

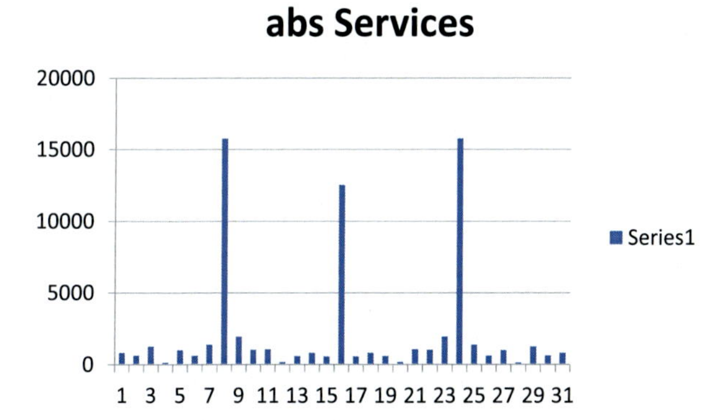

Taxes.

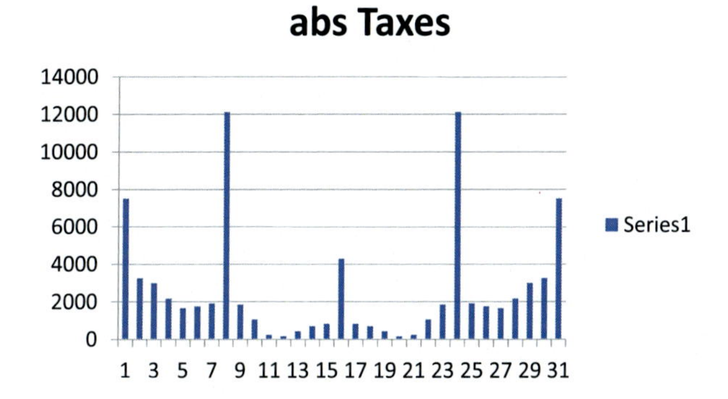

Subsidies.

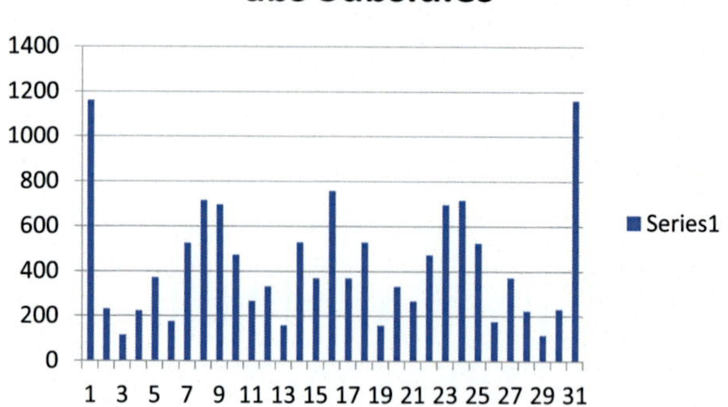

Agriculture.

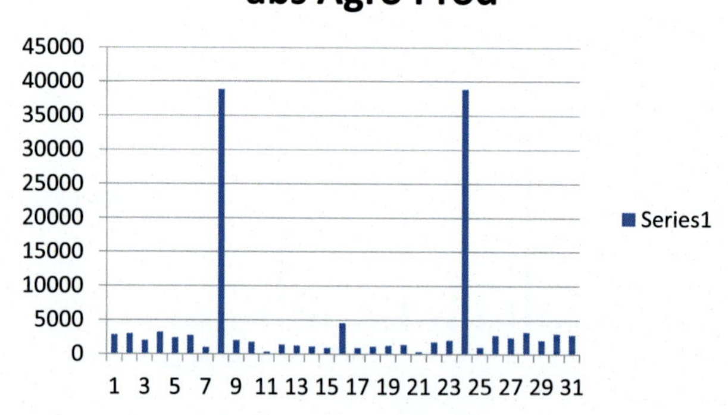

Total GDP.

abs GDP

120000
100000
80000
60000
40000
20000
0
1 3 5 7 9 11 13 15 17 19 21 23 25 27 29 31
Series1

It is important to note that the representations above only give the real values of the amplitudes. The representation in the complex plane is the one that provides the eigenvalues of each component.

Eigenvalues and flows

We are giving below the complex plane representation for the Fourier complex amplitudes. The values are either real or complex conjugate. We see that several main values are greater than all the others. These values determine the dynamic behavior of the processes.

We give the example of the flow of trajectories for some of these components, as determined by the major solutions described above.

The interesting thing to note is the shape of the characteristic potential, on which the trajectories evolve, which is a measure of the speed with which the trajectory converges toward the equilibrium point or diverges from it.

Along this line, we see that the finance sector trajectory slowly diverges from the equilibrium point, while all other sectors converge toward it. Subsidies also have an intricate behavior that should be analyzed in context.

The time constant from the trends above may also be calculated in the cases of step excitation and no excitation as it is shown below.

Industrial production complex plane.

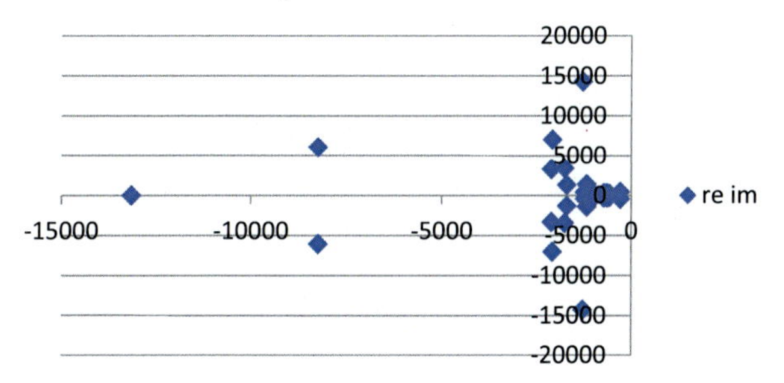

Industrial production flow.

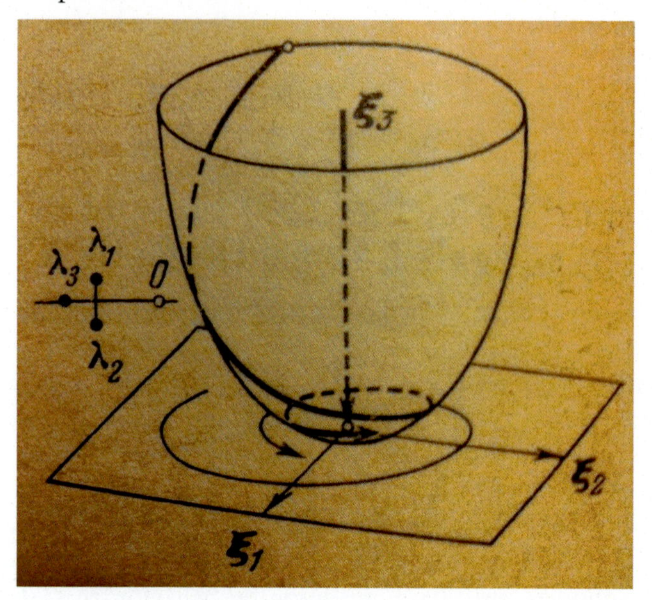

Agriculture complex plane.

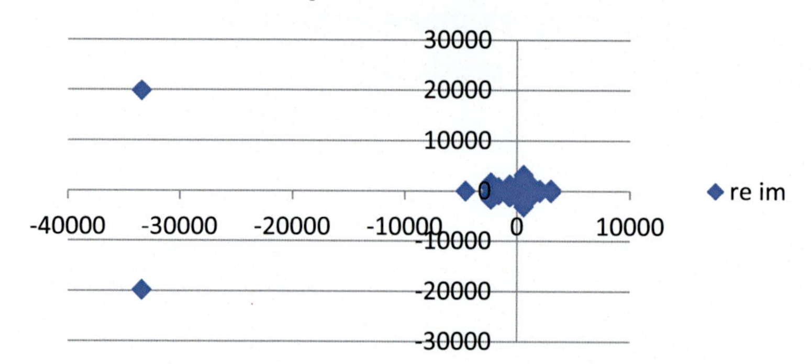

Agriculture flow.

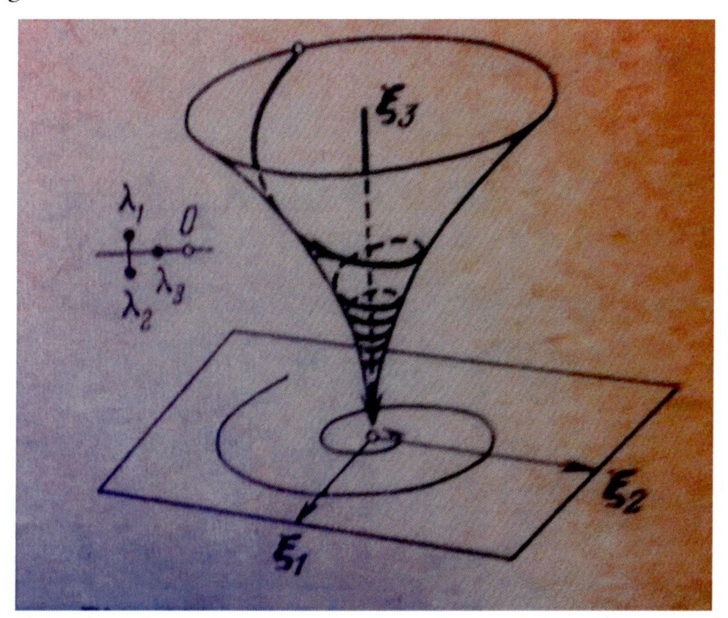

Construction complex plane.

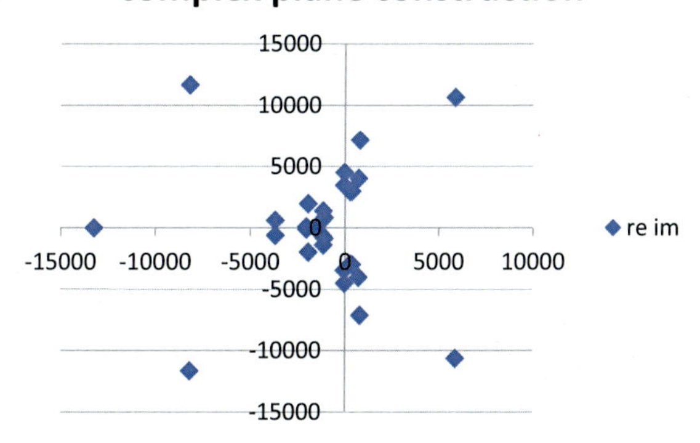

Construction flow.

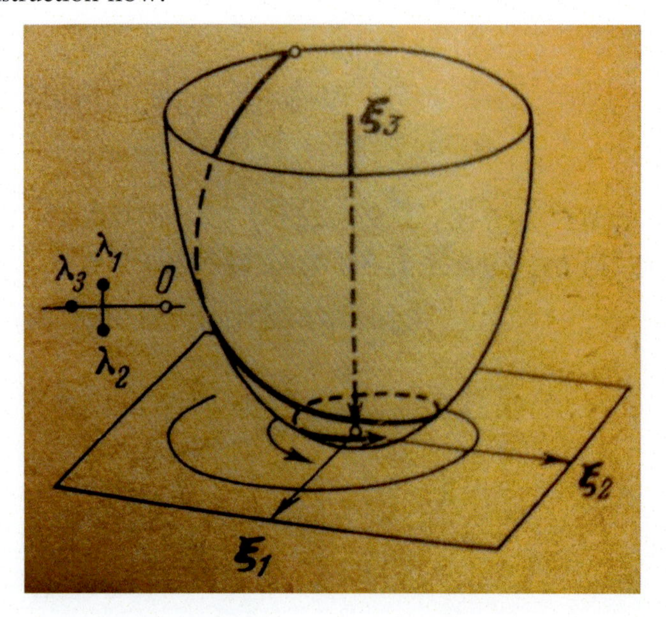

Commerce complex plane.

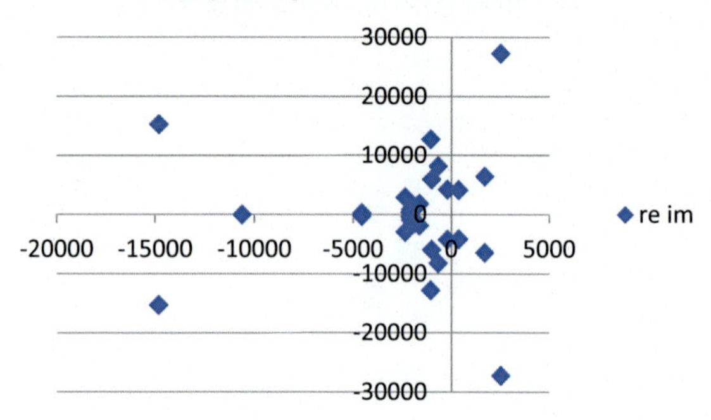

Commerce flow.

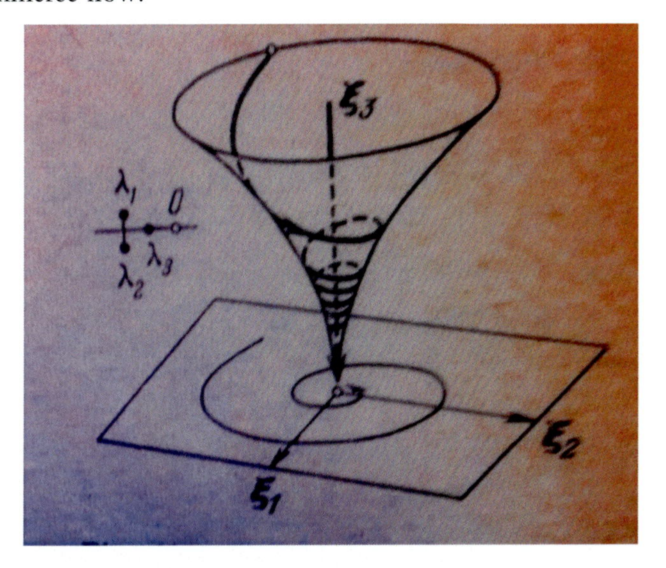

Finance complex plane.

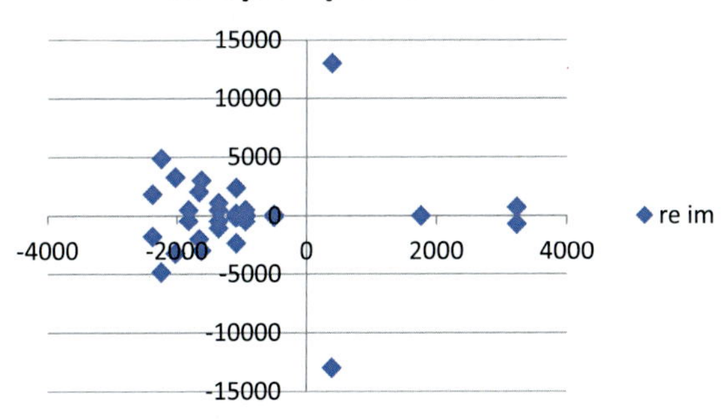

Finance flow.

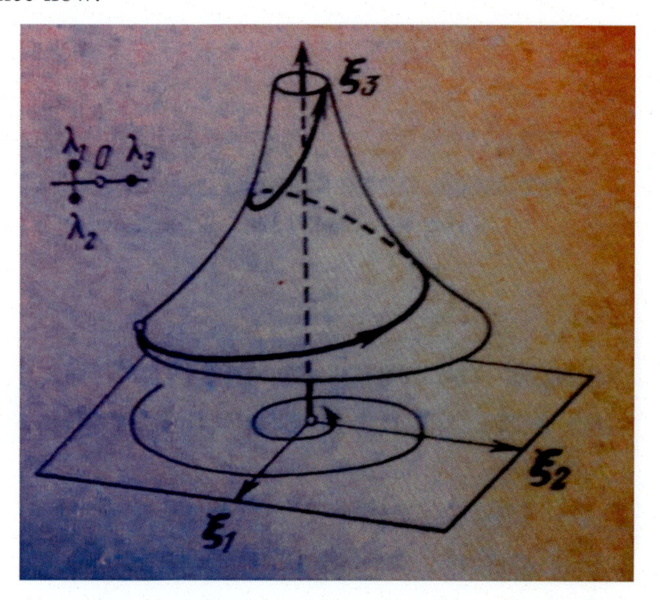

Services complex plane.

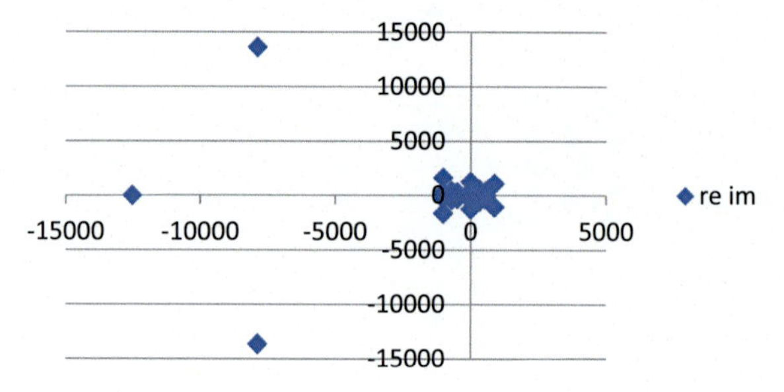

Services flow.

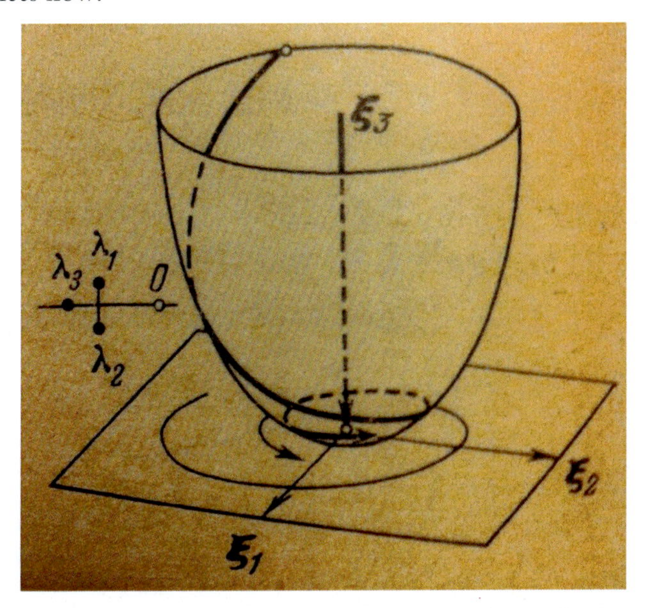

Taxes complex plane.

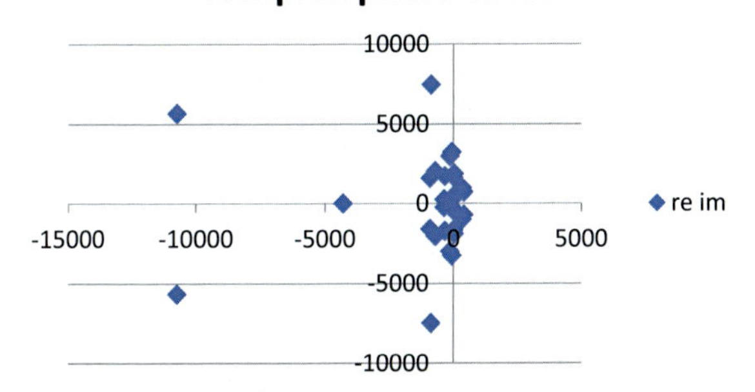

Taxes flow.

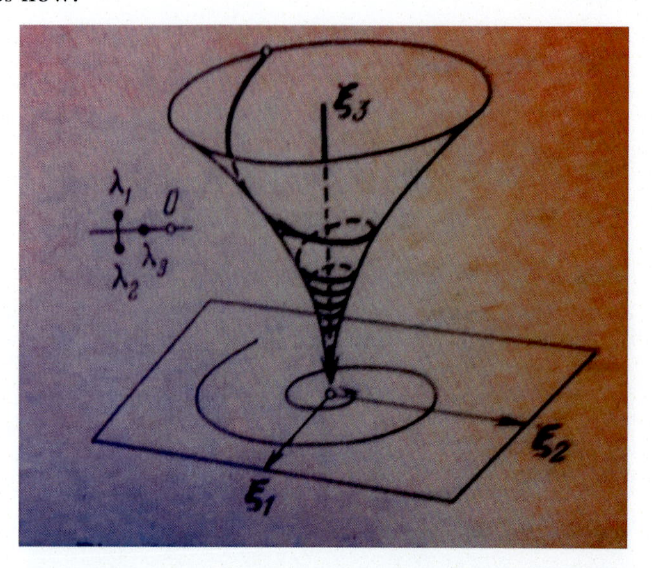

Subsidies complex plane.

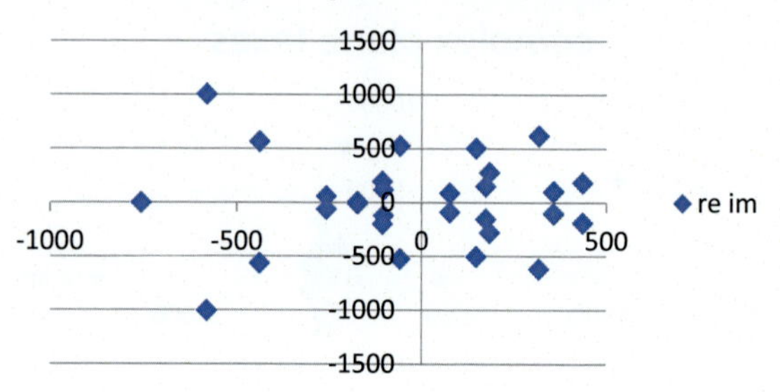

Subsidies flow.

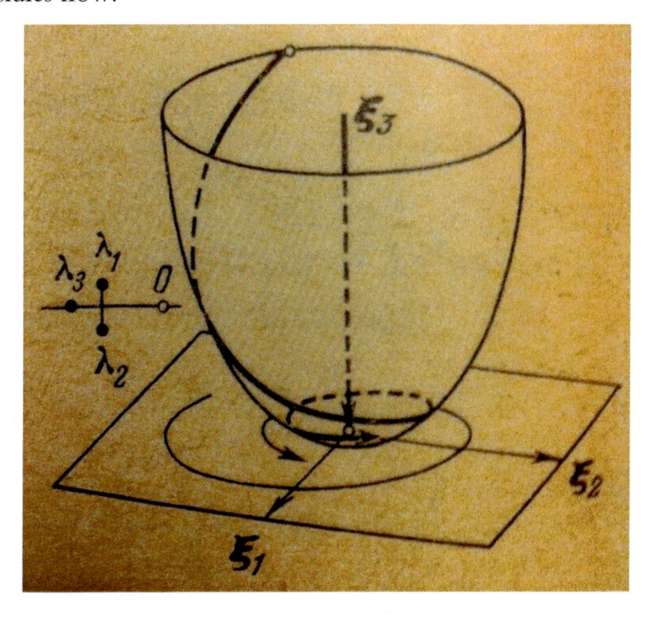

GDP complex plane.

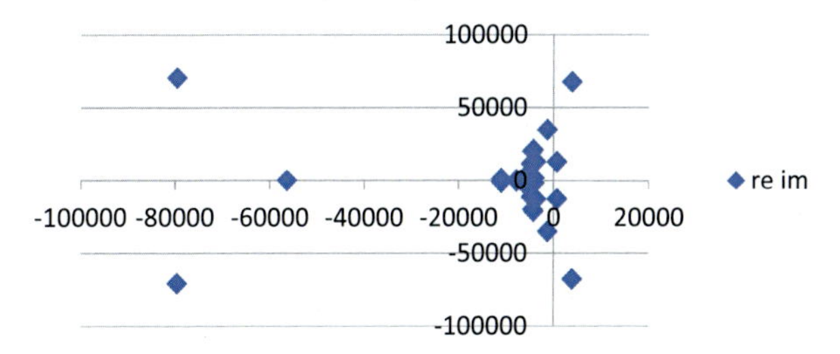

GDP flow.

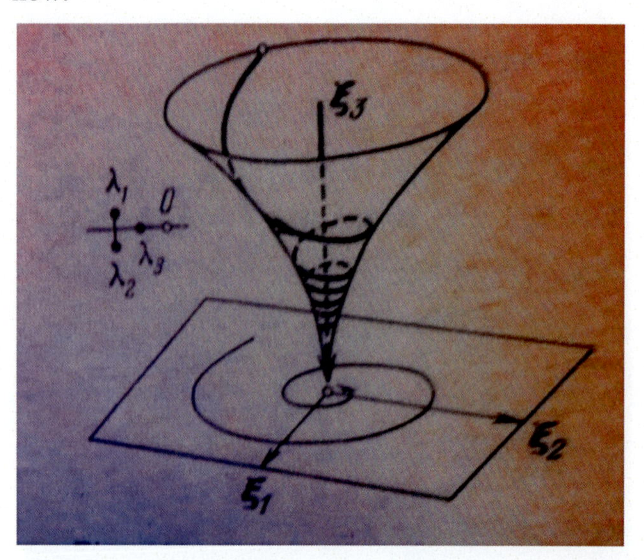

Determination of differential equations

Below, we give the algorithm for determining the coefficients of the second-order differential equation for GDP. Note that we use all 36 values of quarterly data.

The system of second-order differential equations of the GDP components, determined based on the same algorithm as above are listed in Table 5.3, including the natural period T, the dumped period S, and the time constant of reaction C. The last three parameters are measured in quarters. The step function was noted $u(\tau)$.

Table 5.3 Coefficients and dynamic parameters for step function excitation.

Component	M	c	k	T	S	C
AGR	1.026	0.864	1.15	5.934	6.468	2.374
COM	5.452	2.765	5.414	6.305	6.52	3.943
CTR	0.985	0.703	2.428	4.002	4.109	2.802
FIN	1.012	0.824	1.102	6.021	6.538	2.458
IND	2.857	1	3.656	5.554	5.622	5.713
SRV	11.072	6.213	15.907	5.242	5.392	3.564

AGR, Agricultural production; *COM*, Commerce; *CTR*, Construction; *FIN*, Finance; *IND*, Industry; *SRV*, Services.

The equation for each component is:

$$M\frac{d^2}{d\tau^2}x(\tau) + 2\frac{c}{2}\frac{d}{d\tau}x(\tau) + kx(\tau) = u(\tau)$$

The values above give the transient evolution for a step excitation. We have also calculated values for the case of no excitation, which are shown in Table 5.4.

One may note that Table 5.4 provides the period data to calculate both transient and natural frequencies.

Further on we will start analyzing the correlation of various sectors. It is clear that there is mutual interdependence among the sectors, e.g., finance and industry are interdependent as well as finance and every other sector. Moreover, the nonfinancial sectors are interdependent, such as industry and agriculture, agriculture and commerce, and other combinations.

Long- and short-term intersectoral cycles

If one combines two cyclic processes the resulting pattern of behavior results in high frequency cycles (given by the sum of the two frequencies) modulated by lower frequency cycles (given by the difference of the two frequencies). In physics the process is called "beats" and its perception is mostly demonstrated in sound waves.

Until now, we have considered only a single GDP component's oscillatory behavior with no interconnections among sectors. Let us now try to assess more information from combining two sectors and analyze the resulting behavior.

Since the financial sector is behaving capillary in the economy we will consider first the correlation of this sector with each of the others.

We start with the presentation of the basic theory, that is standard in Physics:

Table 5.4 Coefficients for the case of no excitation.

Component	M	c	k	T	S	C
AGR	0.076	12.496	−0.744	−2.01i	−0.077i	0.012
COM	1.744	52.456	−1.303	−7.269i	−0.417i	0.066
CTR	0.469	2.504	−0.141	−11.468i	−2.308i	0.375
FIN	0.771	49.976	−0.64	−6.897i	−0.194i	0.031
IND	0.38	1.652	−0.031	−22.128i	−2.869i	0.46
SRV	0.065	4.994	−0.097	−5.159i	−0.164i	0.026

The resulting wave is a product of the sine of the frequency sum, $\sin((w_f+w_i)/2)$, by the cosine of the frequency difference, $\cos((w_f-w_i)/2)$. The shape of the resulting wave gives information not only on short-term cyclic behavior but also on the modulating cycle that may imply intrinsic long-term cyclic behavior associated with what we usually perceive as "crises" or that have been shown to exist by various authors (of whom the most well known is Kondratieff).

In the case of the finance correlation with each of the other sectors, we provide graphs below of the resulting waves.

We keep the notations from Table 5.4 for the sectors (but in small characters) and give two frequencies for each correlation. We also note that $w = 2\pi/T$ (T from Table 5.4).

Finance and industry (fin ind):

The time units are measured in quarters[Q], so the short period is $2\pi/(\text{wfin}+\text{wind}) = 2.89$ Q, i.e., less than 1 year, while the modulation (long period) is $2\pi/(\text{wfin}-\text{wind}) = 69.78$ Q, i.e., about 17 years.

The period of the modulation is of the order of investment cycles in the industrial sector. We may have found by this method a way to assess the length of the investment cycles in given economies as given by the modulated wave of the correlated financial and industrial cycles of the respective GDP components.

wfin − wind	wfin + wind
−0.09	2.17

The fin−ind frequencies.

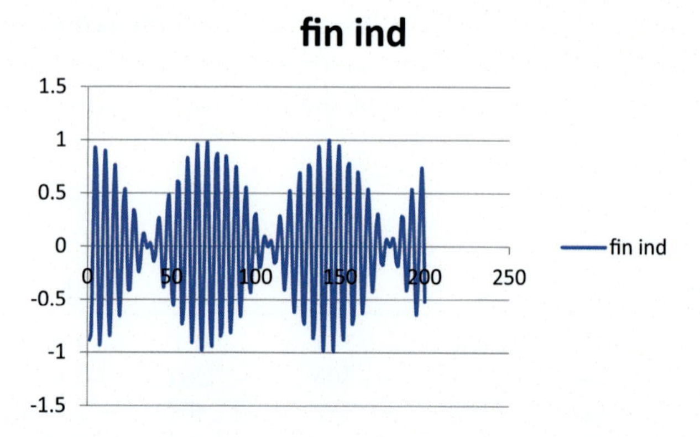

Finance and agriculture

In this case the short-term period is $2\pi/(\text{wfin}+\text{wagr}) = 2.99$ Q, i.e., almost 3 quarters, while the modulation (long period) is $2\pi/(\omega\varphi\iota\nu-\varphi\iota\gamma\rho) = 314$ Θ ι.ε. αβουτ 78 ψεαρσ.

The meaning of this long cycle for Romanian agriculture may represent an interesting research program; we will not insist here on this but go further with the inter-correlations of financial and other sectors cycles.

Fin − agr frequencies

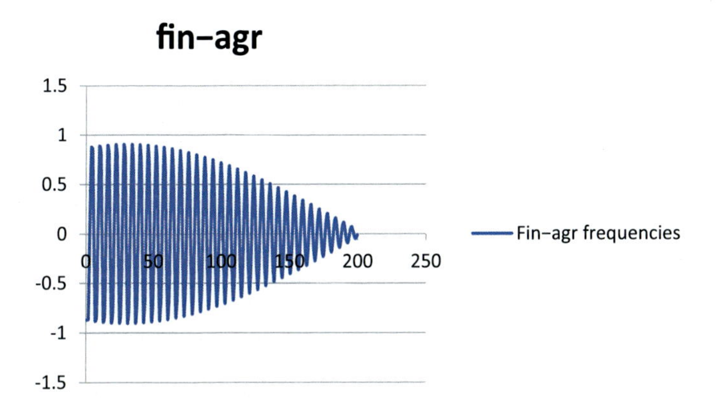

wfin − wagr	wfin + wagr
−0.02	2.10

Finance and commerce

In this case, the short-term period is $2\pi/(\text{wfin}+\text{wcom}) = 3.08$ Q, while the modulation (long period) is $2\pi/(\text{wfin}-\text{wcom}) = 125$ Q, i.e., about 31 years.

The fin−com frequencies.

wfin − wcom	wfin + wcom
0.05	2.04

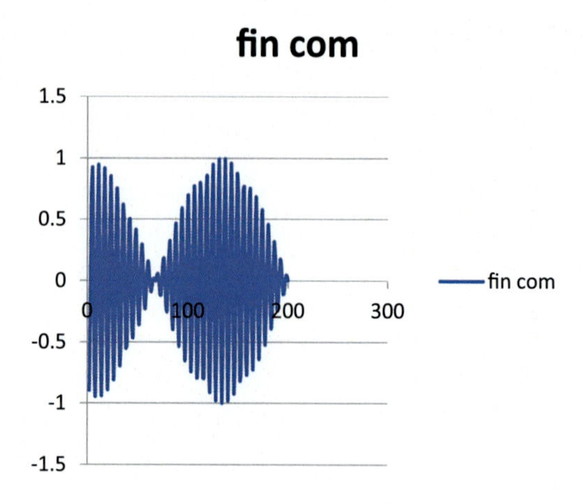

Finance and construction

In this case, the short-term period is $2\pi/(\text{wfin}+\text{wctr}) = 2.4$ Q, while the modulation (long period) is $2\pi/(\text{wfin}-\text{wctr}) = 11.8$ Q, i.e., about 3 years.

As one may see, financial bubbles (modulated wave) in construction are likely to occur over a relatively short time interval, probably indicating caution related to these types of cycles.

The fin–ctr frequencies:

wfin – wctr	wfin + wctr
–0.53	2.61

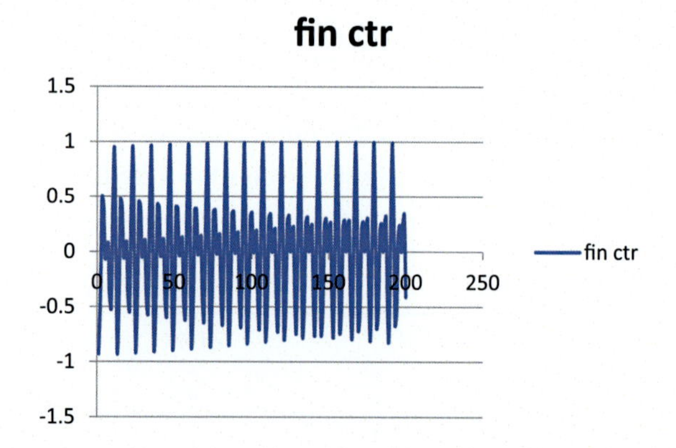

Finance and services

In this case, the short-term period is $2\pi/(\text{wfin}+\text{wsrv}) = 2.8$ Q, while the modulation (long period) is $2\pi/(\text{wfin}-\text{wsrv}) = 39.25$ Q, i.e., about 10 years.

A modulated cycle of 10 years probably indicates that in the Romanian economy, services are still at an incipient level of penetration, since one would have expected a time constant closer to that of the construction sector.

The fin−srv frequencies:

wfin − wsrv	wfin + wsrv
−0.16	2.24

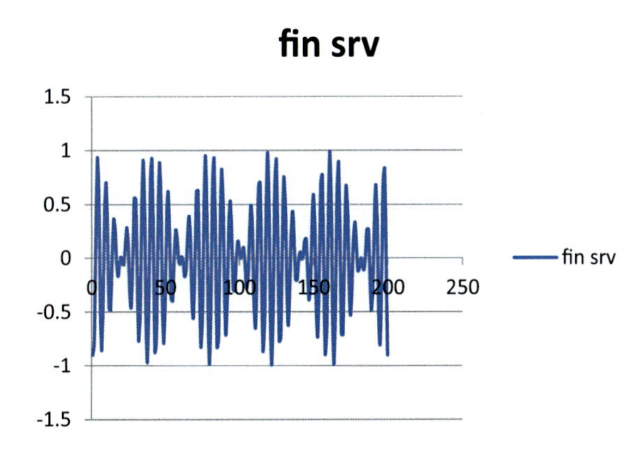

Industry and agriculture

One should note that we are now passing to correlations of nonfinancial sectors such as industry and agriculture. In this case, the short-term period is $2\pi/(\text{wind}+\text{wagr}) = 2.87$ Q, while the modulation (long period) is $2\pi/(\text{wind}-\text{wagr}) = 89.7$ Q, i.e., about 22 years.

Logically, the modulated cycle may represent technological changes in agriculture, but other interpretations of this industry−agriculture correlation are also possible.

The ind−agr frequencies:

wind − wagr	wind + wagr
0.07	2.19

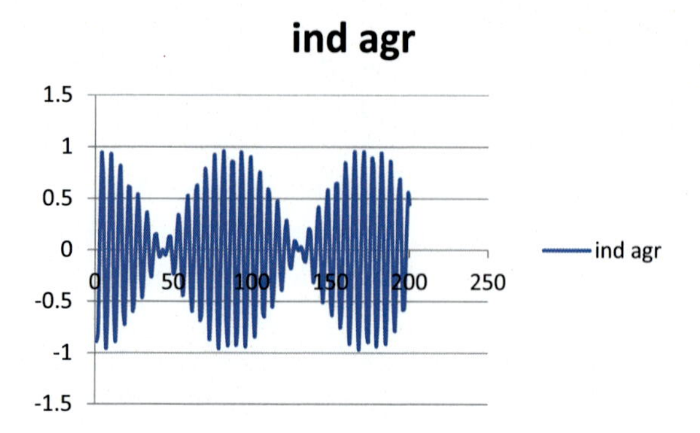

Industry and construction

In this case, the short-term period is 2π/(wind+wctr) = 2.33 Q, while the modulation (long period) is 2π/(wind−wctr) = 14.3 Q, i.e., about 3.6 years.

It is interesting to note that the construction sector is also dynamic in terms of industry involvement, which may explain why an increase in construction activity is perceived as positive by the industrial entities that contribute to screening for the occurrence of bubbles.

The frequencies of ind−ctr:

wind − wctr	wind + wctr
−0.44	2.70

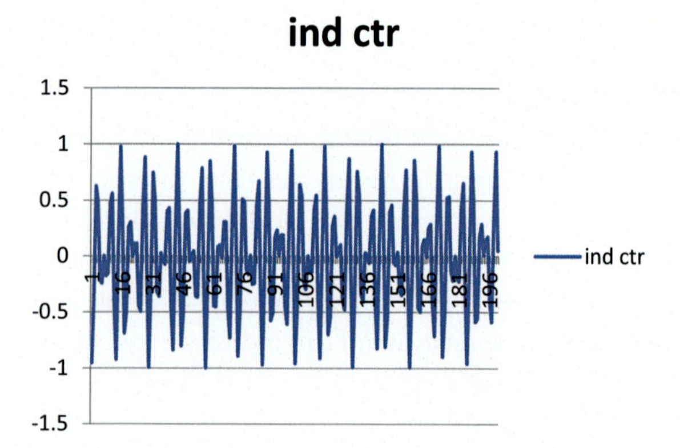

Industry and commerce

In this case, the short-term period is 2π/(wind+wcom) = 2.9 Q, while the modulation (long period) is 2π/(wind−wcom) = 48.3 Q, i.e., about 12 years.

If, as seen above, the modulated cycle of finance and commerce is of the order of 31 years, the 12-year period of industry and commerce shows that selling industrial products is more dynamic in the long run, while short periods are similar among fin–com and ind–com of the order of 3 Q. This probably means that the commerce of industrial products is supported financially to a great extent.

Wind – wcom	wind + wcom
0.13	2.13

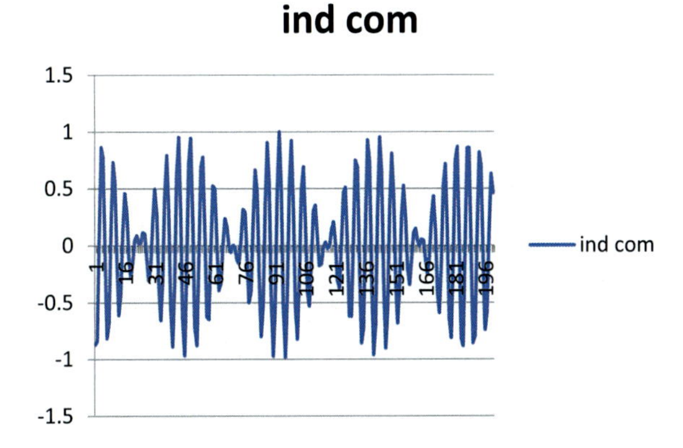

Industry and services

In this case, the short-term period is $2\pi/(\text{wind}+\text{wsrv}) = 2.7$ Q, while the modulation (long period) is $2\pi/(\text{wind}-\text{wsrv}) = 89.7$ Q, i.e., about 22 years.

Although industry and services are dynamic in the short term, the long-term time constant is relatively high, indicating that Romanian industry is likely more oriented toward energy-intensive industries than toward manufacturing for the service sector.

The frequencies of ind–srv:

wind – wsrv	wind + wsrv
−0.07	2.33

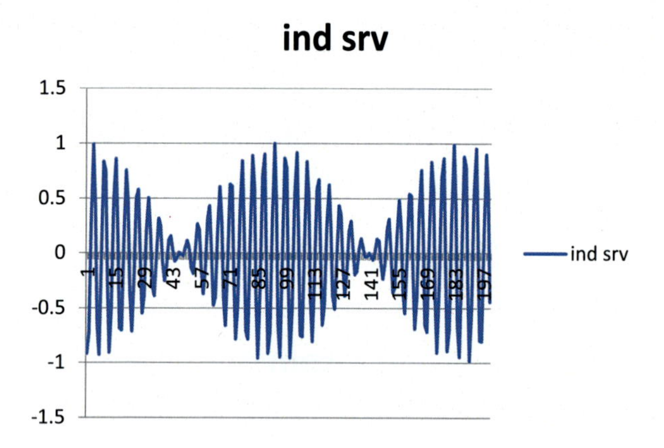

Agriculture and commerce

In this case, the short-term period is 2π/(wagr+wcom) = 3.05 Q, while the modulation (long period) is 2π/(wagr−wcom) = 104 Q, i.e., about 26 years.

The frequencies of agr−com:

wagr − wcom	wagr + wcom
0.06	2.06

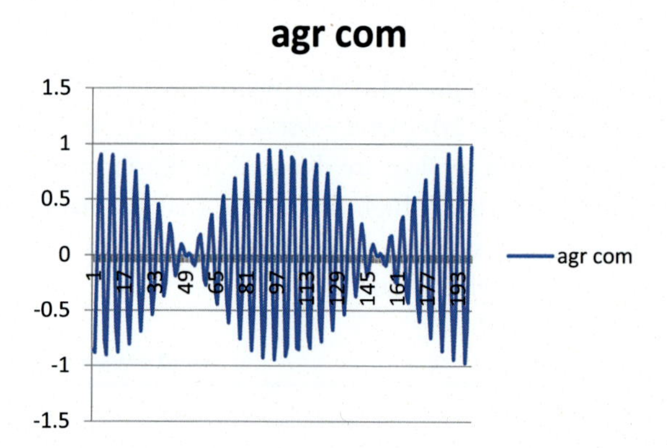

Agriculture and construction

In this case, the short-term period is 2π/(wagr+wctr) = 2.39 Q, while the modulation (long period) is 2π/(wagr−wctr) = 12.3 Q, i.e., about 3 years.

As for other sectors, the correlation of agriculture and construction is dynamic and long term, indicating a large workforce involved in the construction sector.

The frequencies of agr−ctr:

wagr − wctr	wagr + wctr
−0.51	2.63

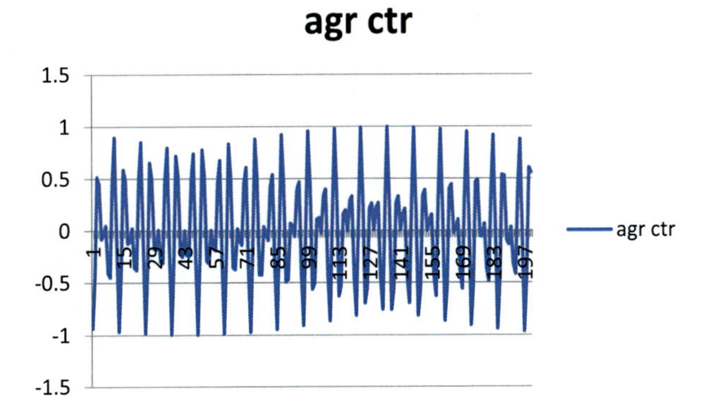

Agriculture and services

In this case, the short-term period is 2π/(wagr+wsrv) = 2.78 Q, while the modulation (long period) is 2π/(wagr−wsrv) = 18.5 Q, i.e., about 4.6 years.

As with construction, agriculture and services have high dynamics, indicating the large number of food-related services.

The frequencies of agr−srv:

wagr − wsrv	wagr + wsrv
−0.14	2.26

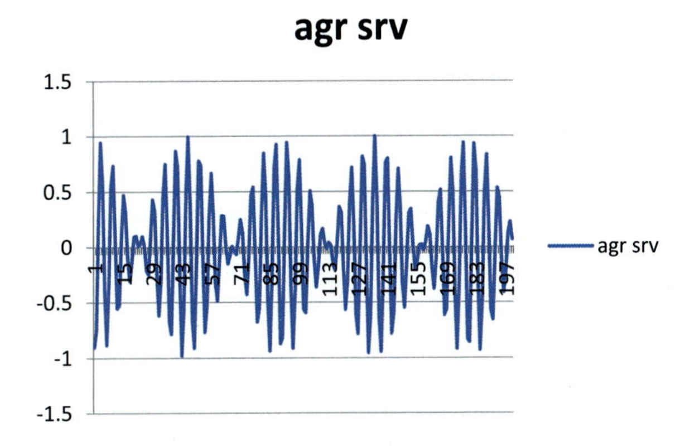

Commerce and construction

In this case, the short-term period is 2π/(wcom+wctr) = 2.4 Q, while the modulation (long period) is 2π/(wcom−wctr) = 11 Q, i.e., about 2.75 years.

As seen above, commerce and construction are the most dynamically intercorrelated even on the modulated wave with a short period.

The frequencies of com−ctr:

wcom − wctr	wcom + wctr
−0.57	2.57

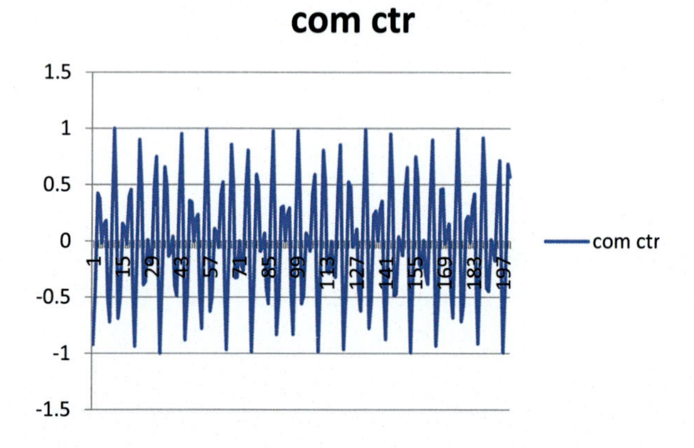

Commerce and services

In this case, the short-term period is 2π/(wcom+wsrv) = 2.85 Q, while the modulation (long period) is 2π/(wcom−wsrv) = 31.4 Q, i.e., about 8 years. If construction and commerce were very dynamic, services and commerce would have a longer time constant for the modulated cycle. In a large service economy, one would have expected to have a shorter time constant, although the short cycle period is rather small.

Frequencies com−srv

wcom − wsrv	wcom + wsrv
−0.20	2.20

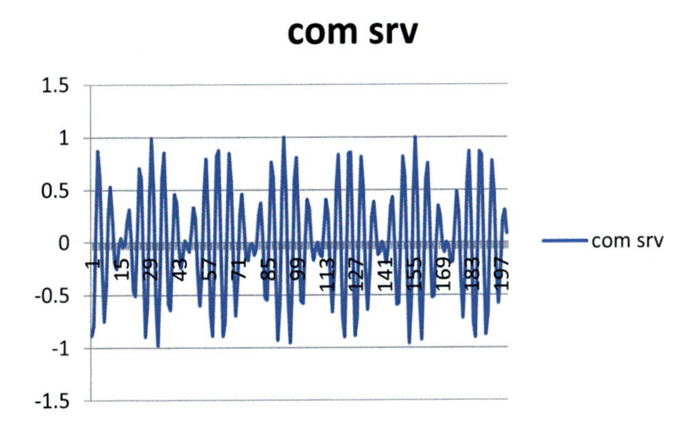

Services and construction

We are now at the end of our correlations by pairs of sectors. In this case, the short-term period is $2\pi/(\text{wsrv}+\text{wctr}) = 2.27$ Q, while the modulation (long period) is $2\pi/(\text{wsrv}-\text{wctr}) = 16.97$ Q, i.e., about 4.24 years.

As for commerce, we witness the high dynamics of services and construction for both short-term and long-term cycles.

Frequencies srv−ctr.

wsrv − wctr	wsrv + wctr
−0.37	2.77

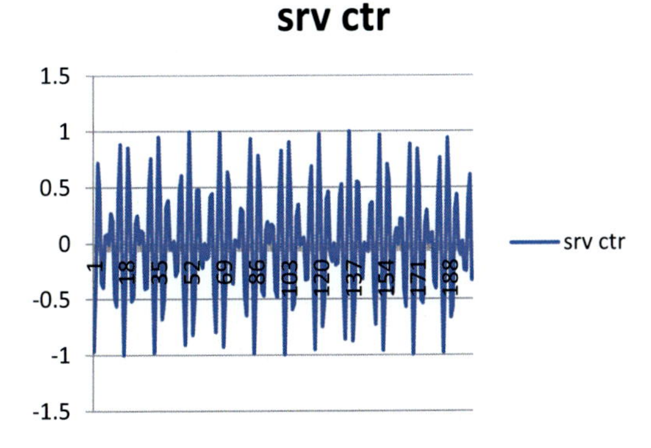

The results above show that an interesting pattern of behavior develops when we consider the possible correlations of pairs of GDP component cycles. The "beatings" process varies from one pair to another, and one may separate short-term behavior from long-term modulated behavior. As the cycle periods are shorter, the dynamics of the components' correlation behavior are more intense.

Modulated cycles are ones with larger time constants that describe behavior covering, in certain cases, tens of years. These long-modulated cycles supposedly generate long-term variations that we normally perceive as crises, being different from the usual quarterly or at most yearly behavior.

Short- and long-term cycle periods may become a set of indicators for the expected behavior; for example, construction correlated with finance, services, and commerce shows similar periodic values for short- and long-term behavior (also relatively small). This high dynamic for these correlations leads to caution related to the construction domain, where a bubble situation may occur. In fact, in the years following the interval we analyzed (2000–08), this situation appeared in Romania.

By contrast, in the case of finance and industry correlations, the difference between short-term and modulated periods is significant, with long-term behavior possibly describing investment cycles associated with industrial development.

Very long-term behavior cycles, like the ones for finance and agriculture, have also been observed, probably indicating that financing dedicated to agriculture should be revised.

From a wider perspective, the graph of added cosine (long-term) modulations for all the correlations shows long-term cycles with periods of 36–41 years. These time constants are similar to Kondratieff cycles spanning very long periods.

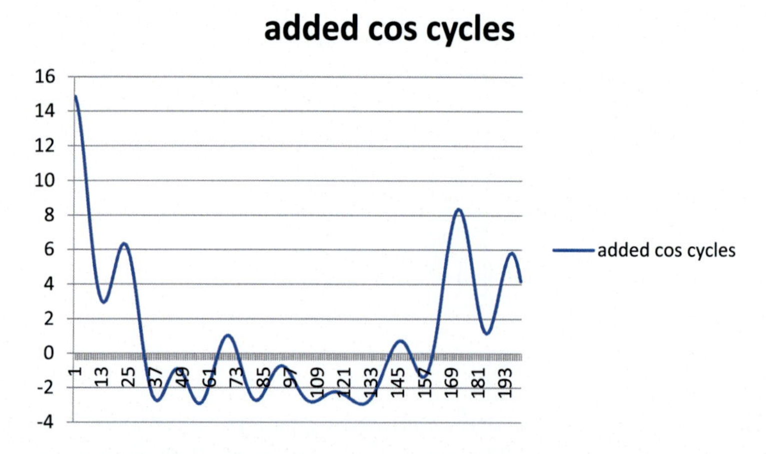

If we consider the added full cycles (sine and cosine products), the long cycles have periods of about 24 years. One may assume here that short-term dynamics have the effect of shortening long-term cycles, i.e., present-day Kondratieff-type cycles may be shorter.

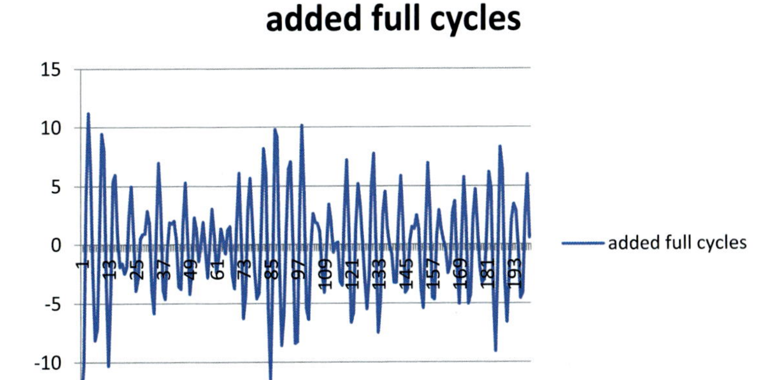

The fact that Kondratiev cycles may be shortened by fast economic cycles is important not only in itself but also relative to the time constant of environmental recovery.

We stress again that development dynamics must consider relations to environmental dynamics if sustainability is to be achieved. The environment does not absorb everything without changing and eliminating the cause of a perturbation. In the future, one of these changes may result in a scenario where 65 million years from now, a civilization of ants may wonder how the former "dinosaurs" on Earth have disappeared.

References

Geng, et al., 2012. Macro-level Circular Economy Indicators. EASAC.

IEA, 2013. Energy Outlook. International Energy Association.

Pearce, D., Markandya, A., Barbier, E., 1990. Blue Print for a Green Economy. Earthscan Publications Ltd., London.

Purica, I., 2010. Nonlinear Models for Economic Decision Processes. Imperial College Press – World Scientific, London, ISBN 978-1-84816-427-7.

Purica, I., 2012a. Nonlinear GDP Dynamics and Basins of Behavior. Lambert Academic Publishing, Saarbrucken, ISBN 978-3-659-29363-4.

Purica, I., 2012b. Oscillatory dynamics of industrial production. Romanian Journal of Economic Forecasting. XV (4), 117–128. ISSN 1582-6163.

Purica, I., 2015. Nonlinear Dynamics of Financial Crises – How to Predict Discontinuous Decisions. Academic Press (Elsevier), London, ISBN 978-0-12-803275-6.

Purica, I., et al., 2016. Report on the Indicators for Circular Economy. EASAC.

CHAPTER SIX

Green investment schemes for sustainability

We must not wait for things to come, believing that they are decided by irrescindable destiny. If we want it, we must do something about it.

Erwin Schrodinger.

This sixth chapter briefly presents various green investment schemes (GISs) designed to help implement the sustainable projects discussed in Chapter 7.

The financing of sustainable projects has gained innovative schemes through the introduction of emission trading certificates. Various international financing institutions have proposed such schemes. What follows are given as examples of schemes from the World Bank (WB) and the Japan Bank for International Cooperation (JBIC), as well as descriptions of typical schemes for general use. We should mention that the period of validity of the Kyoto Protocol, signed in 1997 and valid from 2005 to the Paris Agreement in 2016, has seen the emergence of large numbers of projects based on joint implementation schemes among various countries.

The WB proposed an example GIS for Bulgaria, as presented in a later diagram.

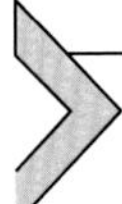

6.1 Case study—green investment scheme of World Bank

The WB proposed a new GIS to the Bulgarian government in October 2004.

The main structure of the proposal:

《First Phase 》

- A GIS fund will be set up under the initiative of the Bulgarian government.
- A portion of the assigned amount units (AAUs) owned by the government will be sold at market and the revenue will be put into the fund.

《Second Phase 》

Climate Change and Circular Economics
ISBN: 978-0-443-29969-8
https://doi.org/10.1016/B978-0-443-29969-8.00007-0

- The fund will provide financial support to the project
  Eligible projects: hard greening and soft greening
  Project selection: fund manager
  Fund manager: external professional fund manager
  Linkage between AAU sales and projects: no direct linkage
  Reserved points on fund-based GIS.
  Potential international buyers seek.
- Real contribution to global warming issues
- Economically effective approach to securing credit
  -Less linkage between AAU sales and projects
- Risk of inefficiency in terms of reduction amount against AAU sales amount

(*AAUs*—assigned amount units—emission reduction certificates at the economy level)

How to improve the linkage between AAU sales and projects

1. AAU transfers project by project during implementation
   ⇒ Project-based GIS
2. Buyer's participation in the decision-making process of the fund
   ⇒ Middle ground between project-based GIS and fund-based GIS
3. Define criteria for eligible projects in a more concrete manner
   Project-based GIS

- To realize the emission reduction project and improve the linkage between AAU sales and actual emission projects
- To simplify the procedure of approval and transfer (similar procedure as JI first track country)
- To set the realistic baseline assumption
- To enhance investment by constructing a reliable structure for potential investors

Structure of Green Investment Scheme (WB's proposal to Bulgaria)

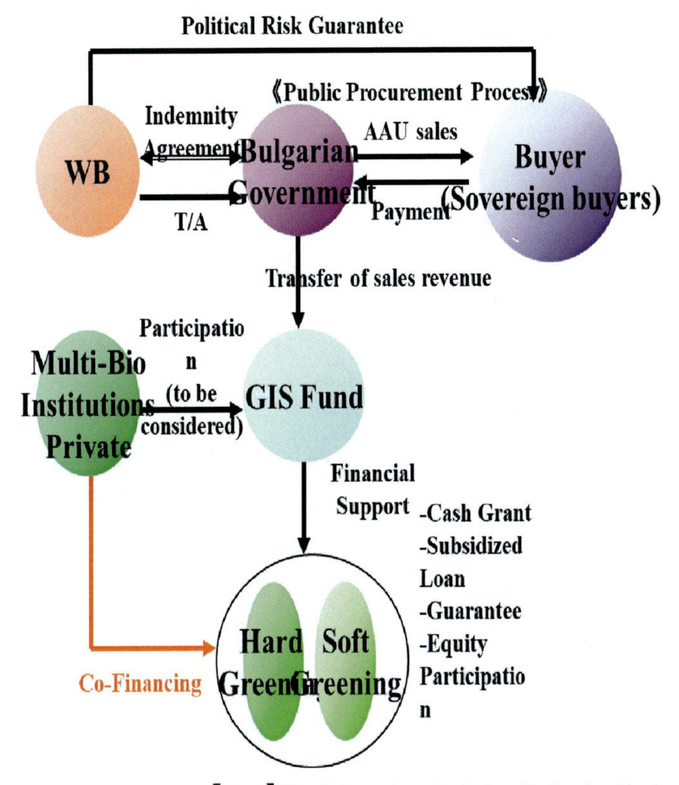

【Note】Hard Greening: Emission Reduction Project
Soft Greening :Other environmental friendly projects, including capacity building

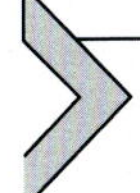

6.2 Case study: Japan Bank for International Cooperation–proposed green investment scheme financing structure

Structure of Financing

Lender: JBIC with commercial banks using export and investment loans
Borrower: GHG Emission Reduction Project: Emission Reduction Project in Bulgaria with Japanese companies' involvement; equity participation, equipment supply, O&M, etc.
Security: Collateral in offshore escrow account (AAU sales revenue by initial allocation)

Emission Reduction: Emission reduction by the project will be part of JBIC loan repayment.

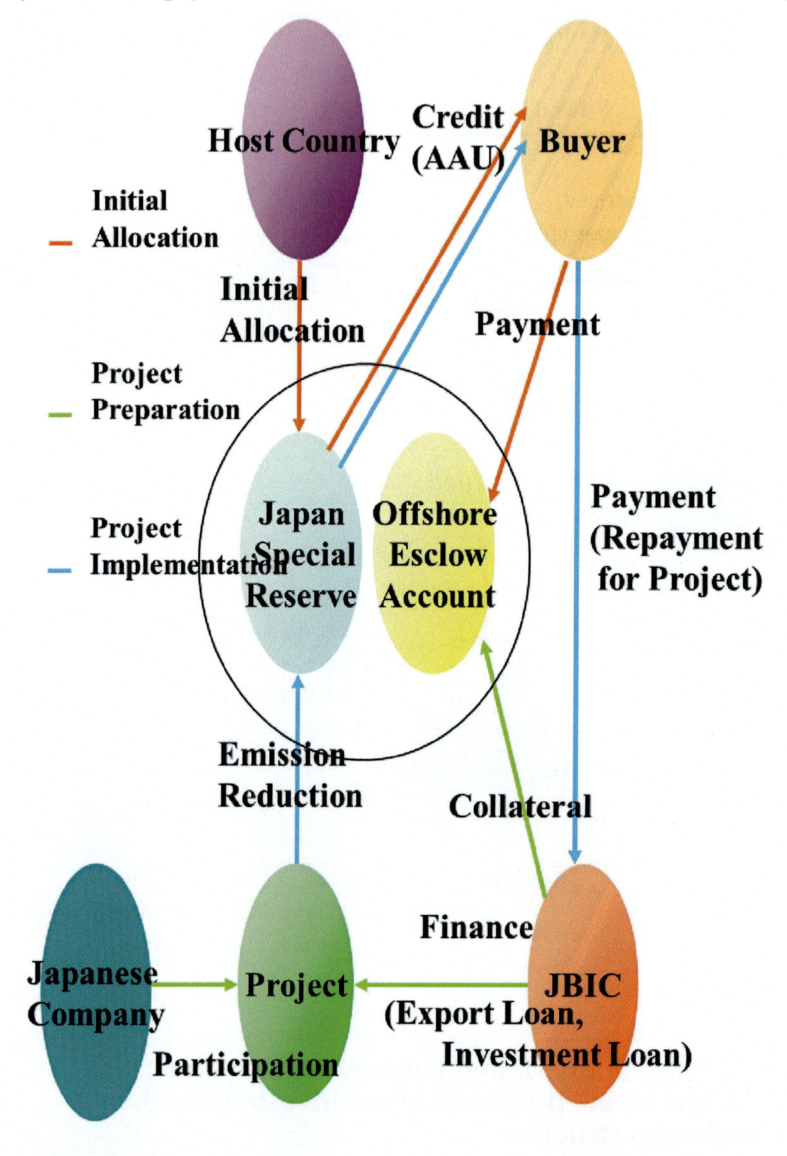

6.3 General green investment scheme for Romania

The diagrams below present two green investment schemes.

The first combines grants with certificate sales (be they AAUs at the economy level or —emission reduction units at the project level) done respectively between governments and companies involved in the investment and project.

The second uses a refinance guarantee fund in combination with the sale of emission certificates.

Legend: *GoRo*—government of Romania; *GoPa*—government of partner country; *RoCo*—Romanian company; *PaCo*—partner company

Various more complex schemes have been used in projects based on emission certificate trading. The ones given here are examples of potential schemes. Financial creativity has high limits when it comes to designing investments for sustainable projects.

Combine with grant schemes

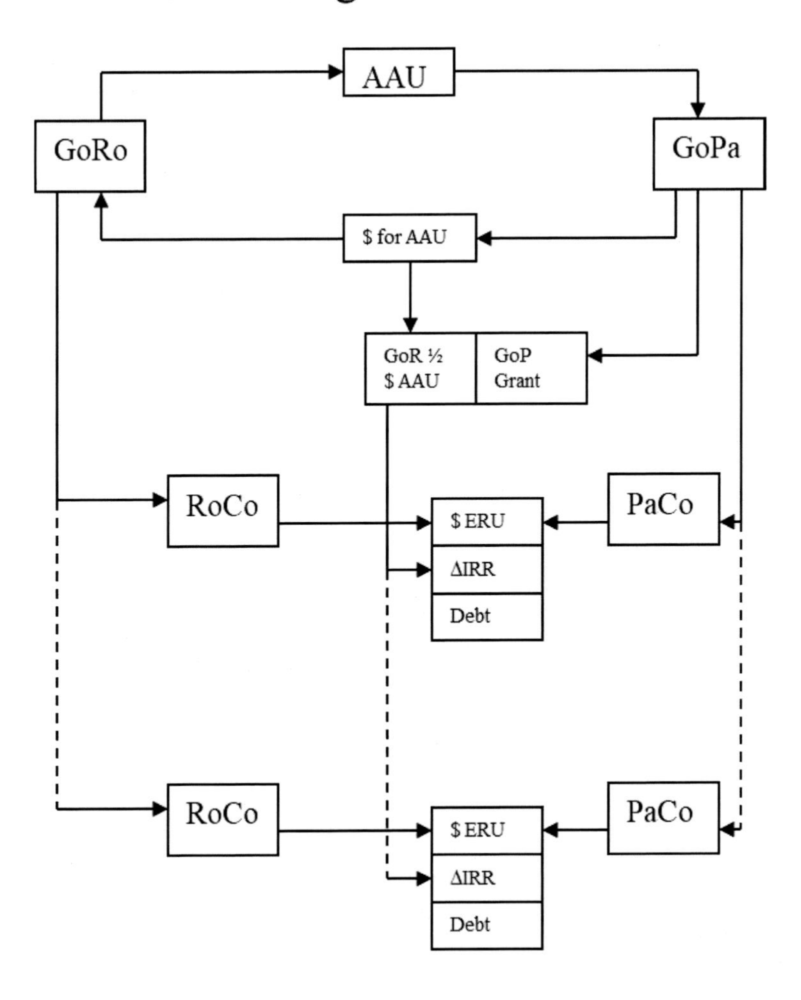

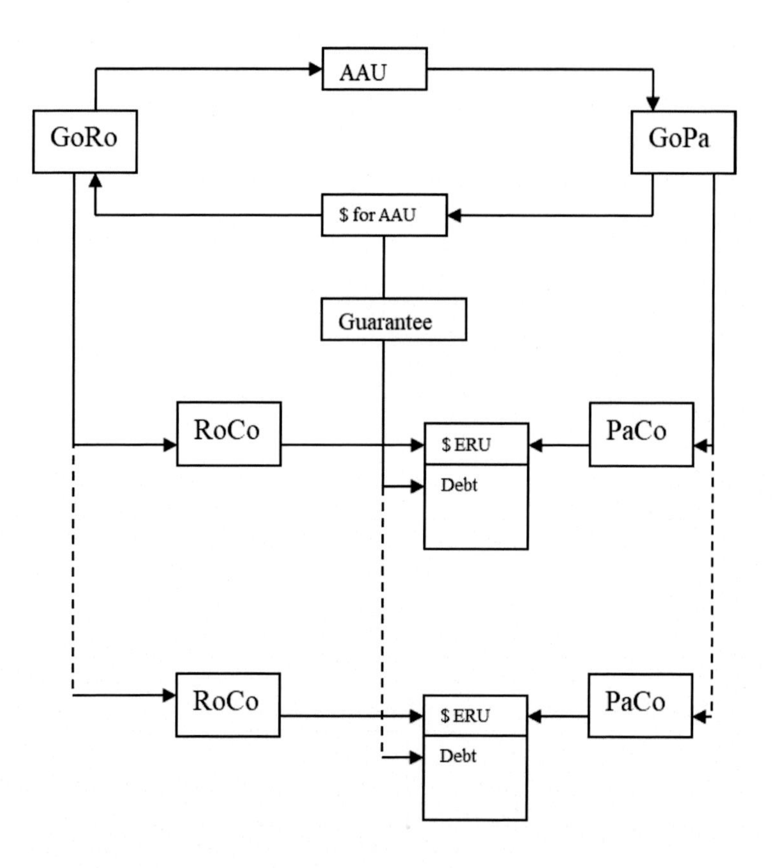

At the Conference of the Parties (COP) 15 in 2009, developed countries committed to a collective goal of mobilizing USD 100 billion per year by 2020 (extended to 2025 at COP 21) to support climate action in developing countries. In 2020, public climate finance for energy, transport, and industry reached USD 31 billion and mobilized USD 9 billion in private capital. In 2020, each USD 1 of public climate finance in the sector leveraged just USD 0.3 of private finance. Average leverage ratios for private capital mobilization must reach USD 6–7 by the early 2030s under the Net Zero Emissions by 2050 Scenario (NZE). We are in front of a trust issue here; for private investment to engage in these projects, risk coverage is needed from governments. The perception of project validity needs a more complex indicator than emission reduction, an indicator that does not reflect financial criteria.

A recent International Energy Agency report (*Net Zero Roadmap A Global Pathway to Keep the 1.5°C Goal in Reach, 2023 Update*) concludes that much more is needed to deliver on the commitment made at COP 15 and provide what is needed under the NZE to put the world on course to global net zero emissions by 2050. Moreover, considerable additional funding will be needed so that state-owned enterprises that cannot presently access commercial capital can fund projects such as electricity network upgrades or energy efficiency improvements in public buildings.

Along the same lines, COP 28 (in 2023) is attempting to mobilize financial resources for net zero project initiatives.

As a general conclusion related to financing sustainable projects, correlation with the environment is crucial in considering investment viability. The best example is the financing of district heating projects. The usual term considered by a bank is measured in months or years. In this case, financing is metaphorically applied to the turning of the Earth around the Sun. One should realize that the business cycle of heating is linked to the winter season. Hence, payback terms should be measured in winters.

Neglecting the relationship of human society with the environment in assessing business cycles and blindly applying financial algorithms may lead to skipping financially viable and sustainable project investments with consequences beyond just financial costs.

The research of entropy in economic processes is complex and covers a tremendous set of topics. In the next chapter, selected texts provide an image of that complexity. We need a unified approach and vision related to how human society may eventually relate to nature as a closed thermodynamic system.

Entropy in economics (bioeconomics, thermoeconomics, econophysics, and others)

The secretary tells her boss that she has repeated a calculation ten times to be sure she has made no mistake ... and presents him with the ten results.

The seventh chapter presents a selection of literature related to entropy in economics to provide a sense of the complexity of the topic. Given the recent use of ChatGPT, I have also included some text generated on the same topic (this text is of the order of magnitude of the other selected texts and should be considered an example.

The discussions that follow include excerpts from the internet related to energy and entropy relative to various economies and economic processes. I selected these excerpts, which are nonexhaustive, to show the diversity of views associated with the topic under discussion. Moreover, there is also an answer from ChatGPT.

Some of the most important and urgent topics requiring economic analysis and policy advice are the problems of climate change and environmental sustainability, and what can be done to alter corporate and individual behavior to deal with these issues. Neoclassical economists tend to focus on market solutions such as carbon trading, drawing on ideas of perfect rationality of actors and the appropriateness of "marginal" analysis. To link such policies to the whole range of potential actions, from legislative and regulatory to changing individual behaviors, requires the economy and society to be analyzed in its full complexity, recognizing that "marginal" analysis can be not just irrelevant but positively harmful when the need is for systemic shifts in economic and social trajectories. This article draws upon a seminar series on complexity economics to consider how heterodox economic analysis can be brought to bear on the issue of the environment, to develop a realistic policy agenda for change. ***Towards a new complexity economics for sustainability*** *Timothy J. Foxon, Jonathan Köhler, Jonathan Michie and Christine Oughton**

The idea that low-entropy matter-energy is the ultimate natural resource requires some explanation. This can be provided easily by a short exposition of the laws of thermodynamics in terms of an apt image borrowed from Georgescu-Roegen.

Climate Change and Circular Economics
ISBN: 978-0-443-29969-8
https://doi.org/10.1016/B978-0-443-29969-8.00010-0

Consider an hour glass. It is a closed system in that no sand enters the glass and none leaves. The amount of sand in the glass is constant—no sand is created or destroyed within the hour glass. This is the analog of the first law of thermodynamics: there is no creation or destruction of matter-energy. Although the quantity of sand in the hour glass is constant, its qualitative distribution is constantly changing: the bottom chamber is filling up and the top chamber becoming empty. This is the analog of the second law, that entropy (bottom-chamber sand) always increases. Sand in the top chamber (low entropy) is capable of doing work by falling, like water at the top of a waterfall. Sand in the bottom chamber (high entropy) has spent its capacity to do work. The hour glass cannot be turned upside down: waste energy cannot be recycled, except by spending more energy to power the recycle than would be reclaimed in the amount recycled. As explained above, we have two sources of the ultimate natural resource, the solar and the terrestrial, and our dependence has shifted from the former toward the latter.

—Daly and Cobb, For the Common Good.

7.1 Storage, emergy, and transformity

"The literature on evaluation of nature is extensive, much of it reporting ways of estimating market values of the storehouses and flows in environmental systems. In some approaches to environmental evaluation, monetary measures were sought for the storages of nature. Others have used the simple physical measures of stored resources, especially energy."

"Shown in Figure 12.1 a is a storage of environmentally generated resources. Energy sources from the left are indicated with the circular symbol. Energies from sources are used in energy transformation processes to produce the quantities stored in the tank. Following the second law, some of the energy is degraded in the process and is shown as "used energy" leaving through the heat sink, incapable of further work. Also due to the second law the stored quantity tends to disperse, losing its concentration. It depreciates, with some of its energy passing down the depreciation pathway and out through the used energy heat sink."

"To build and maintain the storage of available resources, work requiring energy use and transformation must be performed. Work is measured by the energy that is used up, but energy of one kind cannot be regarded as equivalent to energy of another kind. For example, 1 joule of solar energy has a smaller ability to do work then 1 joule of energy contained in coal, since the coal energy is more concentrated than the solar energy. A relationship between solar and coal energy could be calculated by determining the number of joules of solar energy required to produce 1 joule of coal energy. The different kinds of energy on earth are hierarchically organized with many joules of energy of one kind required to generate 1 joule of another type. To evaluate all flows and storages on a common basis, we use solar emergy defined as follows":

> *"Solar emergy is the solar energy availability used up directly and indirectly to make a service or product. Its unit is the solar emjoule."*

"Although energy is conserved according to the first law, according to the second law, the ability of energy to do work is used up and cannot be reused. By definition, solar emergy is only conserved along a pathway of transformations until the ability to do work of the final energy remaining from its sources is used up (usually in interactive feedbacks) (Table 7.1)."

"Solar transformity is defined as follows":

> *"Solar transformity is the solar emergy required to make 1 joule of a service or product. Its unit is solar emjoules per joule."*
>
> ***[pp. 201–203].***

Entropy, Environment and Resources (Second Edition), M. Faber, H. Niemes, and G. Stephan; Springer-Verlag, 1995. Phone: 1-800-Springer. ISBN 3-540-58984-8.

Table 7.1 Emergy of some global storages (natural capital)[a] possible orders of magnitude.

Item	Replacement time, years[b]	Stored emergy, sej	Macroeconomic value, 1992 Em$
World infrastructure[c]	100	9.44 E26	6.3 E14
Freshwaters	200	1.89 E27	1.26 E15
Terrestrial ecosystems	500	4.7 E27	3.1 E15
Cultural and technol. information	1 E4	9.44 E28	6.3 E16
Atmosphere	1 E6	9.44 E30	6.3 E18
Ocean	2 E7	1.89 E32	1.25 E20
Continents	1 E9	9.44 E33	6.3 E21
Genetic information of species	3 E9	2.8 E34	1.86 E22

sej, solar emergy joules.
[a]Product of annual solar energy flux, 9.44 E 24 sej/yr and order of magnitude replacement times in column 1.
[b]See notes in A.
[c]Highways, bridges, pipelines, etc [p. 209].
Investing in Natural Capital, ISBN 1-55963-316-6 published by The International Society for Ecological Economics and Island Press, 1994. Phone: 800-828-1302 or 707-983-6432; FAX: 707-983-6164.

7.1.1 Preface to the second edition

This book has been used as a text in the Department of Economics at the University of Heidelberg for the last decade and the University of Bern (Switzerland) for the last 7 years. We were therefore glad when Dr. Muller of Springer-Verlag offered to publish a softcover version of the second edition to make the text more economically accessible to students.

7.1.2 From content

In Part II we develop our natural science starting point (cf. Fig. 0.1). Since the notion of entropy is very difficult to understand and at the same time of central importance for our approach, we devote the larger part of Chapter 3 to its introduction. It is well known that economics has been strongly influenced by classical mechanics for about a century. The development of thermodynamics since the beginning of the 19th century, however, has remained largely unnoticed by economists. For this reason we have chosen to present the thermodynamic relationships of importance to us in detail. We hope in this way to highlight the differences between classical mechanics and thermodynamics. Thermodynamic processes are irreversible and thus process-dependent with time; Clausius noticed this temporal aspect and introduced the notion of entropy, which stems from the Greek verb "turn over" (turn back, change). It can be argued that it was from classical mechanics that economists derived the attitude that economic processes are fully controllable once they have been fully described. Thus, in many models of growth theory the initial conditions and the growth rate suffice for determining the values of all variables at all times. The study of thermodynamic processes, however, shows that uncontrollable variables are present in addition to those that are controllable. Economists, of course, have noticed this, too. The following remark by Leontief (1953:14) still applies to many economic analyses even today:

> *"In principle at least, it has long been recognized that the ultimate determinants of the structural relationships which govern the operation of the economic system are to be sought outside the narrowly conceived domain of economic science. Notwithstanding their often expressed desire to cooperate with the adjoining disciplines economists have more often than not developed their own brand of psychology, their special versions of sociology, and their particular "laws" of technology.*

It remains to the critics to decide how much this is also true for our Part II. Here, we only wish to mention that Chapter 3 was written for economists and may—except for Sections 3.5, 3.8, and 3.9—be skipped by readers with a natural science background.

In Chapter 4 we use the notion of entropy to establish a relationship between economic activities and the environment. We shall interpret the separation process in the extraction of resources as a reverse diffusion process. Thereafter we shall derive relationships between resource quantities, resource concentration, entropy change, energy, and factor inputs. We shall use these to show how changes in the environment influence the economic production process. We shall establish the relationship between the economic system and the environment as a supplier of resources by way of resource concentration. We thus directly use a variable of nature. With our entropy approach we extend the resource problem beyond the quantitative problem, by the inclusion of aspects of the distribution of resources within the environmental sector and the specific conditions within resource deposit sites.

These two aspects are explicitly considered in Part III, which deals with "The Use of Scarce Resources with Decreasing Resource Concentration." In Chapter 5 we integrate the resource problem into our capital-theoretic approach using the same model structure as in Chapter 2. The common basic model is extended, however, by a resource sector. The waste treatment problem, on the other hand, stays temporarily outside the analysis. We shall, however, consider changes in resource quantities and concentrations within the environmental sector. In Chapter 6—similarly to Chapter 2—we investigate the properties of our model by analyzing the effects of rearranging production on the temporal distribution of the supply of consumption goods. In doing so, we are also interested in the replacement of techniques as a function of resource availability. We then derive optimality conditions for the temporal use of the environment as a supplier of resources. With the help of the variable "resource concentration" we show how the long-run increase in resource extraction costs can be explained as resulting from technological and ecological conditions.

In Part IV we analyze the interdependencies of environmental protection and resource use. For this purpose we join the environmental model of Chapter 2 with the resource model of Part III in a five-sector model.

With examples of resource recoveries from waste materials (recycling) and the controlled deposition of waste materials in the environmental sector, we show how our approach can be used to simultaneously investigate both environmental protection measures and resource use [pp. 6–8].

7.1.3 From the back cover

In this book the authors analyze environmental protection and resource use in a comprehensive framework where not only economic but also natural scientific aspects are considered. With this interdisciplinary procedure, an attempt is made to incorporate the irreversibility of economic processes. The special features of the book are (1) that the authors use a natural scientific variable, entropy, to characterize the economic system and the environment, (2) that environmental protection and resource use are analyzed in combination, and (3) that the replacement of techniques over time is analyzed. A novel aspect of this work is that resource extraction is interpreted as a reverse diffusion process. Thus a relationship between entropy, energy, and resource concentration is established. The authors investigate the use of the environment both as a supplier of resources and as a recipient of pollutants with the help of thermodynamic relationships. The book therefore provides a new set of tools for environmentalists and economists.

Energy and the Ecological Economics of Sustainability, John Peet; Island Press, 1992. ISBN 1-55963-160-0. Phone: 800-828-1302 or 707-983-6432; FAX: 707-983-6164.

"The fifth statement of the second law of thermodynamics is not so obvious as the previous ones, but it brings in some of the points just discussed: In spontaneous processes, concentrations tend to disperse, structure tends to disappear, and order becomes disorder."

"In thermodynamics, there is a concept called entropy that is a measure of the amount of energy no longer capable of conversion into work after a transformation process has taken place. It is thus a measure of the unavailability of energy. Entropy can also be shown to be a measure of the level of disorder of a system. Thus, the pile of broken pieces of china on the floor has a greater entropy than would the same plate, unbroken, on the floor, and the plate on the floor has a greater entropy than did the same plate on the table. If the second law is restated in yet another way, still equivalent to the others, in order to bring in the concept of entropy, it becomes this: All physical processes proceed in such a way that the entropy of the universe increases." [p. 43].

7.2 Information and entropy, by Alan McGowan

Adaptive information (e.g., genetic information) represents stocks of high biological order[1] bought with energy degradation (photosynthesis driving natural selection) over long periods.

In fact, in H. T. Odum's emergy theory of value,[2] genetic information turns out to have far and away the largest emergy, orders of magnitude more than human economic infrastructure. (Emergy is the amount of photosynthetic energy (net primary production) that goes into producing a biological or cultural structure.)

But the people who urge that information can provide a free lunch (technological Cargoists) are not talking about adaptive information, which results from eons of selection within the biological systems hierarchy (coevolution)—or at the very least, from expensive cultural selection processes such as scientific discovery. They are talking about the fact that the cultural noise level is rising—e.g., we have more TV shows and computer games than we used to. They assume that this mostly junk information—which is growing exponentially as the population and economy do—is "just as good" (or even better!) than adaptive information made by evolution, so they think this information somehow promotes our survival. In

[1] Note that a crystal has very low entropy/high order but also low dynamical freedom. Living systems have vast amounts of dynamical freedom in their internal chemical states, but that freedom is kept within the much smaller subset of phase space where biological integrity is maintained. Limits to maintaining integrity are manifestations of the second law in biology. Ecosystem integrity has three sorts of limits:(1) Functional/physiological constraints and tolerances. These include the most basic energetic constraints within life, e.g., the efficiency of photosynthesis, glycolysis, and the rates of DNA replication and repair, mitosis and meiosis, cellular detoxification systems, and other processes. These very ancient bits of frozen historicity were established long before the Cambrian during the evolution of single-celled life. They ultimately set the rates of energy flow and nutrient recycling through ecosystems. They set the objective time of life at its most basic, biochemical level.(2) Evolutionary constraints. i.e., has evolution produced species with the ability to exploit or tolerate a particular set of conditions (and thereby provide a nutrient flow)?(3) Historical constraints. That is, has the historical process (e.g., sweepstakes migrations and local extinctions) that produced the mix of species at a site provided the site with species that can exploit a particular set of conditions?Actually, these are all "historical constraints"—just from quite different historical depths. The functional constraints may come from as far back as the beginning of life. Evolutionary constraints are the adaptive limits of the biota on the current page of geological time in a given geographic region. Meanwhile, historical constraints are those of the assembly of the ecosystem at the site—what species actually wound up there. These constraints limit function and resilience at the site.

[2] See Odum, H.T. 1994. The emergy of natural capital, in A. M. Jansson, M. Hammer, C. Folke, and R. Costanza (eds.). Investing in natural capital: the ecological economics approach to sustainability. Island Press.

other words, they have mistaken growth in mere cultural *variation* for evolutionary adaptation brought about by selection acting on variation. But just as lower-fitness gene variants are not weeded out when a population is released from selection and expands exponentially, junk cultural information is tolerated better under growth conditions, whereas ecologically limited cultures cannot waste excessive human informational bandwith on nonadaptive or maladaptive cultural information.

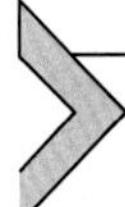

7.3 The entropy concept in biology, by Alan McGowen

- **In ecology**: Ecosystem function (nutrient cycling) arises from the physiologies of organisms. The functions are a network of nutrient pathways (a food web), the *topology* of which is determined by the ecological strategies of the species and is an essentially fixed evolutionary feature on the time scale of ecosystem processes. The *rate of flow* at a point in the network is determined by population growth in the species there, which is constrained by nutrient supply and physiological rates (also essentially fixed by evolution).

Thus, chemical thermodynamics applies, via physiology, to the energetics of ecosystem function. The entropy concept is imported from chemistry, which in turn derives it from physical (statistical) thermodynamics.

- **In evolution**: Here the problem is to explain the energetics of the production of genetic information. Hierarchical information theory (HIT), based on Shannon entropy, is the tool, and the microcanonical ensemble is defined on a space of genetic variation. It is well known that *formal* analogies exist between HIT and statistical thermodynamics. Whether these formal analogies reflect shared abstractions requires further evidence. Brooks and Wiley (1988), argue that in the case of biological information, the shared abstraction analogies are strong enough to accept HIT entropy as physical entropy.
- **The entropy concept in ecological economics**: Ecological economics is not concerned with the time scales of macroevolution. But it should be concerned with those of microevolution, since microevolutionary processes maintain the adaptive potential/resilience of natural capital over time horizons we can reasonably hope will support human societies and economies. (Though not necessarily the *same* societies and economies throughout the whole interval.)

Thus for ecological economics, the ecological (chemical) entropy concept must be supplemented with HIT entropy sufficiently to account for the maintenance of evolutionary potential, including speciation potential, but not enough to account for the production of whole biotas over geologic time. Additionally, suitable economic and cultural forms of "entropy" presumably need to be added to the picture—but the biological systems within which economies function (or fail to be functional) fall within the arena of evolutionary ecology, and this is the domain of the entropy concept into which an economic entropy concept should be fitted.

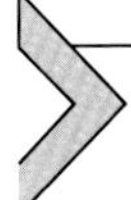

7.4 Microsoft Encarta encyclopedia: second law of thermodynamics

"The second law of thermodynamics gives a precise definition of a property called entropy. Entropy can be thought of as a measure of how close a system is to equilibrium; it can also be thought of as a measure of the disorder in the system. The law states that the entropy—that is, the disorder—of an isolated system can never decrease. Thus, when an isolated system achieves a configuration of maximum entropy, it can no longer undergo change: It has reached equilibrium. Nature, then, seems to "prefer" disorder or chaos. It can be shown that the second law stipulates that, in the absence of work, heat cannot be transferred from a region at a lower temperature to one at a higher temperature."

"The second law poses an additional condition on thermodynamic processes. It is not enough to conserve energy and thus obey the first law. A machine that would deliver work while violating the second law is called a "perpetual-motion machine of the second kind," since, for example, energy could then be continually drawn from a cold environment to do work in a hot environment at no cost. The second law of thermodynamics is sometimes given as a statement that precludes perpetual-motion machines of the second kind."

Is Capitalism Sustainable, O'Conner; The Guilford Press, 1994. ISBN 0-89862-594-7.

7.4.1 Energy and the forces of production

"Bio-economists view economic processes from the point of view of the principles of thermodynamics, insisting that these principles apply both to natural systems and to systems rearranged or transformed by man. The second law of thermodynamics highlights a key aspect of all productive processes: economic activity, intended to satisfy human needs, runs against the general tendency of the universe to move toward a state of greater disorder, of higher entropy. Human labor runs against this tendency toward increasing disorder of the physical world. It sets into motion the energy sleeping within nature, converts "wild" energy into "domesticated," useful energy. But to make this useful energy available, a certain amount of human energy must be expended, either in the form of energy stored in machines or in the form of living human labor."

By definition, the overall increase of entropy associated with any process of production is always greater than the local decrease of entropy corresponding to this process. Economic activity therefore does not escape the laws of physics: the organizational status of the economy only increases insofar as that of the universe as a whole decreases. As Nicholas Georgescu-Roegen observes, "In the perspective of entropy, every action of a human or of an organism, and even every process of nature, can lead only to a deficit for the overall system." "Not only," he continues, "does the entropy of the environment increase with every liter of gasoline in the tank of your car, but a substantial part of the free energy contained in the gasoline, instead of driving your car, will be reflected in a further increase of entropy .... When we produce a copper sheet from copper ore, we reduce the disorder entropy of the ore, but only at the price of a further increase of entropy in the rest of the universe." Living beings too are subject to this law. Every living organism, including human beings, strives to maintain its own entropy at a constant level by drawing low-entropy energy from its environment, particularly in the form of food. According to Georgescu-Roegen, "In terms of entropy, the cost of any economic or biological undertaking is always greater than the product, in such a way that activities are necessarily reflected in thermodynamic deficit."

"Labor is not, all on its own, the primary self-renewing power conceived by Marxist theory. Its reproduction depends totally on a continuous input of low-entropy energy. This energy derives from the sun directly (rays, heat) or indirectly (wind, hydraulics), from solar radiation stored in fossil fuels (oil, coal, gas), and, in small part, from geothermal flows and nuclear energy. Energy cannot be created by labor or machines: it is always drawn from the environment. Even this extraction is governed by certain constraints. Just as labor is necessary to produce labor, energy is necessary to extract energy from the environment. And just as in a growth economy labor can produce more than what is necessary for its own reproduction, so the energy extracted from nature is generally greater than the energy expended for its extraction. The ratio of labor obtained to labor expended is a critical magnitude in economics: it is imperative that it be greater than 1. Similarly, the surplus corresponding to the difference between energy obtained and energy invested is net energy." [pp. 42—43].

Valuing the Earth, Daly and Townsend; MIT Press, 1993. ISBN 0-262-54068-1800-356-0343 or 617-253-2884.

"Erwin Schrodinger (1945) has described life as a system in steady-state thermodynamic disequilibrium that maintains its constant distance from equilibrium (death) by feeding on low entropy from its environment—that is, by exchanging high-entropy outputs for low-entropy inputs. The same statement would hold verbatim as a physical description of our economic process. A corollary of this statement is that an organism cannot live in a medium of its own waste products." [p. 253].

For the Common Good, Daly and Cobb; Beacon Press, 1989. Phone: 800-631-8571; FAX: 617-723-3097; ISBN 0-8070-4703-1.

7.4.2 Entropy

"Reflection on the use of energy leads immediately to the second law of thermodynamics. The law asserts that entropy is increased when work is done. The notion of entropy is often misunderstood, so it requires a brief explanation."

"The first law of thermodynamics declares that energy (or matter-energy) can neither be created nor destroyed. This seems to suggest that the use of energy will not reduce the amount of energy available to be used again. But this is not the case. The second law declares that whenever work is done, whenever energy is used, the amount of useable energy declines. The decline of useable energy is the increase of entropy (the increase of sand in the bottom chamber of the hour glass, to recall the analogy in the Introduction). For example, when a piece of coal is burned, the energy in the coal is transformed into heat and ash. This, too, is energy, and the amount of energy in the heat and ashes equals that previously in the coal. But now it is dispersed. The dispersed heat cannot be used again in the way it was originally used. Furthermore, any procedure for reconcentrating this energy would use more energy than it could regenerate. In other words, the dispersal of previously concentrated energy would increase. There is no way of reversing this process. Burning a piece of coal changes the low-entropy natural resource into high-entropy forms capable of much less work. In spite of the circular flow celebrated by economists there is something that is irrevocably used up, namely capacity for rearrangement. The economic process (production followed by consumption) is entropic. Raw materials from nature are equal in quantity to the waste materials ultimately returned to nature. But there is a qualitative difference between the equal quantities of raw and waste material. Entropy is the physical measure of that qualitative difference. It is the quality of low entropy that makes matter-energy receptive to the imprint human knowledge and purpose. High-entropy matter-energy displays resistance and implasticity. We cannot with any currently imaginable technology power a steamship with the heat contained in the ocean, immense though that amount of heat is. Nor can windmills be made of sand or ashes."

"When nature and its resources for human use are viewed as concentrations of useable energy instead of as passive matter, it will no longer be possible to ignore the fund-flow model of Nicholas Georgescu-Roegen, to whom we owe the path-breaking analysis of The Entropy Law and the Economic Process (1971), which we have freely drawn from."

"Georgescu-Roegen's fund-flow model begins with the recognition that nature's contribution is a flow of low-entropy natural resources. These raw materials are transformed by a fund of agents (laborers and capital equipment), which do not themselves become physically embodied in the product. Labor and capital funds constitute the efficient cause of wealth, and natural resources are the material cause. Labor and capital funds are "worn out" and replaced over long periods of time. Resource flows are "used up" or rather transformed into products over short periods of time. While there may be significant substitutability between the two funds, labor and capital, or among various resource flows, for example aluminum for copper or coal for natural gas, there is very little substitutability between funds and flows. You can build the same house with fewer carpenters and more power saws, but no amount of carpenters and power saws will allow you to reduce very much the amount of lumber and nails. Of course one can use brick rather than wood, but that is the substitution of one resource flow for another rather than the substitution of a fund for a flow. Funds and flows, efficient and material causes, are complements, not substitutes, in the process of production."

"From this commonsense perspective it is very difficult to understand the current neoclassical models of production which (a) often do not include resources at all, depicting production as a function of labor and capital only; (b) if they do include resources, assume that "capital is a near perfect substitute for land and other natural resources"; and (c) fail to recognize any physical balance constraint, that is, do not rule out cases where output constitutes a greater mass than the sum of the masses of all inputs (which would be a violation of the First Law of Thermodynamics). Some recognition of the last problem exists and some efforts have been made to limit substitution by a mass balance constraint on production functions. Economists are occasionally embarrassed by their infractions of the first law, but their more egregious violations of the second law have induced very little shame so far."

"Georgescu-Roegen argues that all resources, and indeed all items of value, are characterized by low entropy; but not all items characterized by low entropy have economic value. Value cannot be explained in only physical terms, but neither can it be explained purely in psychic terms of utility without reference to entropy, as neoclassical economics attempts to do. Since we neither create nor destroy matter-energy it is clear that what we live on is the qualitative difference between natural resources and waste, that is, the increase in entropy. We can do a better or worse job of sifting this low entropy through our technological sieves so as to extract more or less want satisfaction from it, but without that entropic flow from nature there is no possibility of production. Low-entropy matter-energy is a necessary but not sufficient condition for value. It is critically important, therefore, to analyze the sources of low entropy (the physical common denominator of usefulness), and their patterns of scarcity."

"As noted in the Introduction we basically have two sources of low entropy: the solar and the terrestrial. They differ significantly in their patterns of scarcity. The solar source is practically unlimited in its stock dimension, but is strictly limited in its flow rate of arrival to earth. The terrestrial source (minerals and fossil fuels) is strictly limited in its stock dimension, but can be used at a flow rate of our own choosing, within wide limits. Industrialism represents a shift away from major dependence on the stock-abundant solar source toward major dependence on the stock-scarce terrestrial source in order to take advantage of the variable (expandable) rate of flow at which we can use it. On the basis of this elementary consideration alone, it was possible for Georgescu-Roegen to predict, back in the 1960s when most economists were talking about feeding the world with petroleum, that exactly the opposite substitution would happen: we would be fueling our cars with alcohol from food crops that gather current sunshine. In Brazil this has already happened. *Homo sapiens* brasiliensis has entered into direct competition with Mechanistra automobilica for a place in the sun. Sugar cane for fuel is displacing rice and beans for food."

"Returning to the issue of the substitutability of capital for resources, our approach is to consider the amount of capital needed in two scenarios: a world of extensive resource depletion and high capital accumulation versus a world of conserved resources and reduced capital accumulation. It is evident that more is needed in a world in which renewable resources have become scarce. Food may be produced hydroponically, but this requires far more capital than producing the same amount of food in naturally fertile soil. Note that here we are speaking of substitution of humanly created capital stock for natural capital stock (soil), and not the substitution of capital for a resource flow. A carrot produced hydroponically embodies just as much matter and energy as one grown in the garden. The extra humanly created capital in hydroponics is not merely a matter of direct costs of equipment, chemicals, and water. It also involves supplying water that will be more expensive than at present. Deforestation will reduce stream flow, increase flooding, hasten the silting of dams, and speed up aquifer depletion. Capital will then be needed for flood control, new dams, diversion of distant rivers, and desalinization of ocean water." [pp. 194–197].[3]

Vision 2020, Laszlo; Gordon and Breach, 1994, 212-206-8900. ISBN 2-88124-612-5.

"The third possible category is that in which systems are far from thermal and chemical equilibrium. Such systems are nonlinear and pass through indeterminate phases. They do not tend toward minimum free energy and maximum specific entropy but amplify certain fluctuations and evolve toward a new dynamic regime that is radically different from stationary states at or near equilibrium."

"Prima facie the evolution of systems in the far-from-equilibrium state appears to contradict the famous Second Law of Thermodynamics. How can systems actually increase their level of complexity and organization, and become more energetic? The Second Law states that in any isolated system organization and structure tend to disappear, to be replaced by uniformity and randomness. Contemporary scientists know that evolving systems are not isolated, and thus that the Second Law does not fully describe what takes place in them—more precisely, between them and their environment. Systems in the third category are always and necessarily open systems, so that change of entropy within them is not determined uniquely by irreversible internal processes. Internal processes within them do obey the Second

[3] See also **Steady-State Economics**, Daly; Island Press, 1991. ISBN 1-55963-071-X.

Law: free energy, once expanded, is unavailable to perform further work. But energy available to perform further work can be "imported" by open systems from their environment: there can be a transport of free energy—or negative entropy—across the system boundaries. [4] When the two quantities—the free energy within the system, and the free energy transported across the system boundaries from the environment—balance and offset each other, the system is in a steady (i.e., in a stationary) state. As in a dynamic environment the two terms seldom balance each other for any extended period of time, in the real-world systems are at best "metastable": they tend to fluctuate around the states that define their steady states, rather than settle into them without further variation."

Our Ecological Footprint, Wackernagel and Rees; New Society Pub., 1996; ISBN 0-86571-312-X Phone: 800-253-3605.

"The Second Law of Thermodynamics (the "entropy law") states that the entropy of an isolated system always increases. This means that the system spontaneously runs down. All available energy is used up, all concentrations of matter are evenly dissipated, all gradients disappear. Eventually, there is no potential for further useful work—the system becomes totally degraded and "disordered." This has significant implications for sustainability":

"Non-isolated systems (such as the human body or the economy) are subject to the same forces of entropic decay as are isolated ones. This means that they must constantly import high-grade energy and material from the outside, and export degraded energy and matter to the outside, to maintain their internal order and integrity. For all practical purposes, this energy and material "throughput" is unidirectional and irreversible."

Modern formulations of the Second Law therefore argue that all highly-ordered, far-from-equilibrium, complex systems necessarily develop and

[4] Change in the entropy of systems is defined by the well-known Prigogine equation $dS = djS + deS$. Here, dS is the total change in system entropy, while djS is the entropy change produced by irreversible processes within it, and deS is the entropy transported across system boundaries. In an isolated system, dS is always positive for it is uniquely determined by djS, which necessarily grows as the system performs work. However, in an open system, deS can offset the entropy produced within the system and may even exceed it. Thus dS in an open system need not be positive; it can be zero or negative. The open system can remain in a stationary state ($dS = 0$) or grow more complex ($dS < 0$). Entropy change in such a system is given by the equation $deS - djS < 0$); that is, the entropy produced by irreversible processes within the system is shifted into the environment [pp. 106—107].

grow (increase their internal order) "at the expense of increasing disorder at higher levels in the systems hierarchy."[5]

"The human economy is one such highly-ordered, complex, dynamic system. It is also an open sub-system of a materially closed, non-growing ecosphere, i.e., the economy is contained by the ecosphere. Thus the economy is dependent for its maintenance, growth and development on the production of low entropy energy/matter (essergy) by the ecosphere and on the waste assimilation capacity of the ecosphere."

"This means that beyond a certain point, the continuous growth of the economy (i.e., the increase in human populations and the accumulation of manufactured capital) can be purchased only at the expense of increasing disorder (entropy) in the ecosphere."

"This occurs when consumption by the economy exceeds production in nature and is manifested through the accelerating depletion of natural capital, reduced biodiversity, air/water/land pollution, atmospheric change, etc."

Natural Capital and Human Economic Survival, Thomas Prugh with Robert Costanza, John H. Cumberland, Herman Daly, Robert Goodland and Richard B. Norgaard; The International Society for Ecological Economics, 1995. Distributed by Chelsea Green Publishing Company, Phone: 800-639-4099 or 603-448-0317; FAX: 603-448-2576. ISBN 1-887490-01-9.

Things fall apart; the center cannot hold; Mere anarchy is loosed upon the world ….

—W.B. Yeats, The Second Coming.

Things fall apart because it is a law—the second law of thermodynamics. (The first law of thermodynamics is the law of conservation, which says that matter and energy cannot be created or destroyed, only transformed. Matter is itself a form of energy, as is shown by Einstein's famous equation, $E = mc^2$.) The second law was developed in connection with steam engines in 1824 by French physicist Sadi Carnot. Carnot realized that using energy to do work (move matter through space) depended on the machine's temperature gradient, i.e., the difference between the hottest and coolest parts. As work is performed, it reduces the temperature differences. Although the

[5] E. Schneider and J. Kay. 1992. Life as a Manifestation of the Second Law of Thermodynamics. Preprint from: Advances in Mathematics and Computers in Medicine (Waterloo, Ont.: University of Waterloo Faculty of Environmental Studies, Working Paper Series) [p. 43].

energy total remains constant, it becomes less available to do further work (Boulding, 1981a).

More generally, using energy makes it less available. The latent chemical energy in fireplace logs is highly available until it is released by burning. Thereafter, although the amount of energy in the heat, gases, and ashes is the same as the amount that was in the wood, it is scattered (unavailable). In theory, it is possible to reassemble the components and reconcentrate the energy, but doing so would take more energy than it would yield (Daly and Cobb, 1989).

Another way to express the entropy law is that in an isolated system, objects and subsystems tend to disintegrate over time. They break, break down, break up, rust, die, decay, wear out, or generally move from a state of higher organization to one of lower organization, from order to disorder. As far as is known, this process always moves in the same direction. (Since entropy is a measure of the disorder in a system, a highly organized system is said to have low entropy, while a disordered system is said to have high entropy. Entropy increases as order decreases.) The breaking down and wearing out of a system or object can be stopped if it is an open system capable of receiving inputs of matter and energy—maintenance—from outside. Even a closed system, which allows only inputs and outputs of energy, can maintain order over time. In an isolated system (one in which there are no inputs or outputs, i.e., no throughput), disorder must increase.

Is life an exception? Doesn't life create order out of disorder? Anyone with small children will immediately doubt this. Yet life appears to be an example of movement from a state of lower organization to one of higher organization. Human beings, for example, gradually grow out of utter helplessness to relative independence, learning to survive in the world and do things of extraordinary complexity, up to and including writing novels, symphonies, and arcane mathematical tracts on things like entropy. In fact, doesn't evolution in general, with its vast, eons-long procession of movement from creatures of one-celled simplicity to blue whales, prove that entropy can be beaten?

Yes and no. On a local scale, yes—life has indeed evolved marvels of increasing organizational complexity. But in terms of the big picture, no. Living creatures exist only by "import" highly complex, low-entropy matter (i.e., eating food), extracting useful energy and materials from it, and "export" wastes of much lower complexity (higher entropy). All life on Earth recycles itself through the ecosphere in this manner, each creature using something from its surroundings (usually including other creatures) to sustain and recreate itself. Matter is not created or destroyed, only broken apart and reassembled to be used again in some other form. As physicist Erwin Schrodinger once put it, life (and evolution) can be seen as the segregation of entropy: "[T]he device by which an organism maintains itself stationary at a fairly high level of orderliness (= a fairly low level of entropy) really consists in continually sucking orderliness from its environment" (Schrodinger, 1967, p. 79). Life creates pockets of order at the cost of disorder elsewhere. Evolution is pollution (Boulding, 1981a,b).

Humans and other living things are thus clearly open systems. However, the biosphere and the Earth itself are closed systems. Matter is essentially constant; little comes into the system except the occasional stray chunk of comet or meteorite, and little goes out except space probes. But in terms of energy, the flow of solar radiation coming in (balanced by the flow of reradiated heat) is continuous and crucial. It is the ultimate answer to the question, How does the economy (and the world) work? Hazel Henderson (1981) tells of a paper delivered by English Nobelist Frederick Soddy in 1921, in which he used the steam locomotive as a metaphor, asking "What makes it go?":

> *In one sense or another the credit for the achievement may be claimed by the so-called engine-driver, the guard, the signalman, the manager, the capitalist, or the shareholder—or, again, by the scientific pioneers who discovered the nature of fire, by the inventors who harnessed it, by Labor, which built the railway and the train. The fact remains that all of them by their united efforts could not drive the train. The real engine driver is the coal. So, in the present state of science, the answer to the question how men live, or how anything lives, or how inanimate nature lives, in the senses in which we speak of the life of a waterfall or of any other manifestation of continued liveliness, is, with few and unimportant exceptions, BY SUNSHINE (p. 225).*

"Needless to say," Henderson writes, "Soddy was considered a crank." But he was right: the steady imports of solar energy drive the life processes of Earth. If the Earth were closed to the solar flow, which is low-entropy energy made generally available to the biosphere through photosynthesis, all life would eventually cease. Of course, the sun is not exempt from the entropy law; it is slowly running down as it burns up its nuclear fuel and will come apart, spectacularly, in a few billion years.

What has entropy got to do with economics?

The laws of thermodynamics are relevant to the economy because economic activity is entropic. Natural resources (low-entropy matter energy) are gathered, processed to separate the useful parts from the rest, manufactured into goods, and transported to the point of sale. Wastes are produced and energy is used up (and made less available) every step of the way. The quantity of raw materials is equal to the quantity of wastes (plus the products, which eventually become wastes), but the two amounts are qualitatively different. The difference is measured in terms of entropy. Economic production is utterly dependent on the availability of low-entropy inputs (Daly and Cobb, 1989).

These inputs come from two sources. As noted above, one is the sun. The other is the Earth, which yields useful minerals, plant and animal life, and fossil fuels. There are obvious differences between the nature of solar and terrestrial inputs, but perhaps even more important is the radical difference in their availability. Solar inputs are essentially unlimited (at least on time scales relevant to human beings), but they flow to the Earth in a comparative trickle, on the order of 100–200 W per square meter of the Earth's surface. Earthly stocks of renewable resources are in turn limited by the availability of solar energy. Earthly stocks of nonrenewable resources (especially fossil fuels, which are deposits of solar energy laid down millions of years ago) are finite, but through technology, we can extract them from the ground and pour them into the economy at enormous rates (Daly and Cobb, 1989).

Technology has enabled the human economy to temporarily suspend its dependence on the low-flow solar source of low-entropy inputs—which is what marked the threshold between preindustrial and industrial society. But technology cannot abolish the entropy law [pp. 42–45].

A Survey of Ecological Economics, Krishnan, Harris and Goodwin; Island Press, 1995.

This truly excellent book has an entire chapter on entropy and how it relates to economics. Here are two selections:

Summary of Recycling, Thermodynamics, and Environmental Thrift by R Stephen Berry.

Published in *Bulletin of the Atomic Scientists 28* (May 1972): 8–15. From the *Bulletin of the Atomic Scientists.* ~ 1972 by the Educational Foundation for Nuclear Science, 6042 South Kimbark, Chicago, IL 60637, USA. A 1-year subscription to the Bulletin is $30.

As environmental considerations become more important in policy decisions and planning, a compelling need has emerged for reliable and robust indices of environmental use. This is particularly true when choosing between alternative policies, which requires the identification of variables that can be quantified, that are general enough to allow comparison between quite different sorts of processes, that provide key measures or indices, and that yield true measures of the amount of use of the environment. Toward this end, the quantities derived from thermodynamics are the most obvious and natural, and they meet all of these criteria.

Thermodynamic potential is a fundamental measure of a system's capacity to perform work. The science of thermodynamics enables us to determine the minimum expenditure of thermodynamic potential to achieve a given physical change. Since every process requires the consumption of some thermodynamic potential, we can compare different processes and select that which is the most thermodynamically efficient. The change in thermodynamic potential associated with a process will measure all of the energy exchanged as well as the effects upon the degree of disorder or dilution, i.e., the entropy of the system.

The two essential forms of stored potential are energy and order. There are multiple forms of energy storage, including hydroelectric facilities, fossil fuels, solar energy, and nuclear technologies. Order is used when, for example, we obtain materials from concentrated ore bodies rather than by finding them distributed evenly over the planet's surface. Some forms of stored potential are readily accessible, while others require considerable effort and energy expenditure before they can be used. Measuring the total stored potential can be quite difficult and involves a considerable amount of guesswork. However, it is possible to measure accurately the change in potential associated with different processes, so that the thriftiest process can be identified and adopted.

This approach is different in practice from the money-based "least cost" method of optimizing production, so it is important to stress the differences between economic and thermodynamic analysis. Economic analysis is based upon perceptions of present value and scarcity as expressed in the marketplace, where the supply and demand framework is modeled on an instantaneous evaluation of the popular perception of shortages. However, "one cannot take seriously using a short-term market analysis to decide, say, in the year 2171, whether all the remaining fossil fuel should be reserved for the chemical industry." [9] But if economists were to determine their estimates of shortage by undertaking increasingly long-term analyses, even with discounting, their estimates would come closer and closer to those made by thermodynamicists. In a sufficiently long time frame, it becomes evident that the most important scarcity is of thermodynamic potential; thus thermodynamic analysis becomes essential.

7.4.3 System defined

Our system is one in which goods manufacturing consumes materials and other resources from the environment. To calculate the real thermodynamic cost of a manufactured object, we evaluate the amount of thermodynamic potential extracted from the environment to produce the good and then subtract the amount of thermodynamic potential that remains stored within the object. In the unrealizable, idealized thermodynamic limit, the thermodynamic potential that resides within an object is identical to the potential extracted from the environment, the net change in potential is zero, and the process has merely transformed one form of potential into another. However, this naive ideal can never be in practice; the net costs are always greater than zero, and there is always a loss in potential both in producing the good and in discarding it. This net loss from production is a true loss, as it cannot be recovered.

7.4.4 Thermodynamic estimates

As an example of this thermodynamics-based approach, the thermodynamics associated with automobile manufacturing can be examined. Specifically, we can consider the amount of thermodynamic potential consumed in mining and manufacturing from "new" raw materials, the amount consumed in recycling processes, and the minimum requirements for an ideally efficient process. The criterion used is one of "thermodynamic thrift," i.e., the idea that it is desirable to minimize the consumption of thermodynamic potential in achieving any particular goal. There are three policies to consider in this regard: (1) maximizing recycling, (2) extending the useful life of goods, and (3) developing more thermodynamically efficient processes for producing the goods in the first instance.

Each step of the manufacturing process involves the transformation of matter from one state to another, via transformation processes that include mining and smelting, manufacturing, normal use, recycling, junking, and natural degradation. Through numerous, complex calculations, actual figures for loss of thermodynamic potential have been calculated in units of total kilowatt hours (kWh) per automobile. An estimate of 5000–6525 kWh per automobile emerges. The estimate of the ideal thermodynamic potential requirement for producing an automobile, on the other hand, is only about 30 kWh.

The enormous magnitude of the gap between actual and ideal thermodynamic potential costs is striking. From this it is evident that our current manufacturing and mining processes "are reflections of the historically developed means of production and transport, rather than of the thermodynamic requirements for creating the ordered structure of an operable machine." [12] The staggering inefficiency manifested in these figures implies the existence of possibilities for vast savings in thermodynamic potential. Even modest improvements in productive processes could generate savings of thousands of kWh per vehicle.

The potential savings from the alternative policy approaches of recycling or extending product life are smaller but significant. Recycling might save between zero and a little over 1000 kWh per vehicle at best. A limitation of these savings from recycling is the need of new car manufacture for some new materials, mostly to maintain the strength of the vehicles, so the savings figures should be halved. Furthermore, even these savings may not be realizable with current recycling technologies. This assessment could change, however, with improved recycling technologies or an increase in the energy costs of mining and smelting.

The savings associated with an extension of the useful life of a product—for example, through enhanced precision in the manufacturing process itself, or improved maintenance procedures—are somewhat harder to quantify. It is certain, however, that the increased costs of more durable manufacture would be somewhat less than the costs associated with the manufacture of a new product. Doubling or tripling the useful life of an automobile could reduce the overall manufacturing costs by perhaps 1000 kWh, and when the reduced mining and smelting needs are factored in, the net savings increase to 2750–4500 kWh per vehicle.

These figures provide a compelling picture of the differences between these three choices: given current technologies, recycling provides small savings at best when compared with those associated with extending product life, which in turn are small compared with the possible savings from new technologies. However, while it is clear which policy would maximize thermodynamic thrift, the relative ease of adopting one policy over another must also be considered. A policy to encourage maximum recycling would require a relatively small perturbation of existing processes. The extension of useful product life, however, would be more difficult, as it requires a change in both manufacturing techniques and consumer attitudes. The basic technologies to implement the ideal system likely do not yet exist, and the costs of developing and especially of implementing them will be very large indeed. However, the potential savings from their development are so vast that the costs will be insignificant in comparison. For example, it is estimated that saving 1000 kWh per vehicle would equal the output of 8–10 large power generation facilities.

It is clear from the example of automobile manufacturing that a policy of thermodynamic thrift ought to be pursued as a national goal. A three-stage course seems desirable: encourage recycling, develop extended-life machines, and pursue the longer-term goal of developing technologies that would operate with efficiencies closer to ideal limits. However, the policy implications of this last and most crucial goal are at odds with much of current federal policy. We should include in the training of scientists and engineers a specific orientation to conducting this type of research. We should also direct public funds and effort into the development of these technologies since, like military and space technologies, the requisite scale of development is too vast for the private sector [pp. 194–197].

7.5 Summary of energy and the US economy

A Biophysical Perspective by Cutler J. Cleveland, Robert Costanza, Charles A.S. Hall, and Robert Kaufmann.

Originally published in *Science 225* (August 31, 1984): 890–897. C 1984 by the AAAS.

Between the mid-1940s and the early 1970s, the US economy showed generally good performance. Since 1973, however, performance indicators such as labor productivity, inflation, and growth rates have been relatively disappointing, and mainstream economic models cannot entirely explain this shift and its underlying causes. A theoretical perspective that recognizes the importance of natural resources, especially fuel energy, may help; some economic problems can be understood more clearly by explicitly accounting for the physical constraints imposed on economic production.

In this perspective, the focus is on the production process, i.e., the economic process that upgrades the organizational state of matter into lower entropy goods and services. This process involves a unidirectional, one-time throughput of low-entropy fuel that is eventually lost as waste heat. Production is a work process, and like any work process, it will depend on the availability of free energy. The quality of natural resources is also important to this process because lower-quality resources will always require more work to upgrade them into final goods and services.

Based on this biophysical perspective, four hypotheses are presented and discussed below.

7.5.1 Energy and economic production

Hypothesis 1: A strong link between fuel use and economic output exists and will continue to exist.

Rather than viewing the economy as a closed system, it must be seen as an open system embedded within a larger global system that depends on solar energy. The global system produces environmental services, foodstuffs, and fossil and atomic fuels, all of which are derived from solar and radiation energies in conjunction with other important resources. Fossil and other fuels are used by the human economy to empower labor and to produce capital. Fuel, capital, and labor are then used to upgrade natural resources to produce goods and services. Production is a process using energy to add order to matter. Since fuels differ in the amount of economic work they can do per unit of heat equivalent, both the quantity and the quality of fuel play a role in determining levels of economic production. An important quality of fuels is the amount of energy required to locate, extract, and refine the fuel to a socially useful state. This can be measured by a fuel's energy return on investment (EROI), which is the ratio of the gross fuel extracted to the economic energy required directly and indirectly to deliver the fuel in a useful form.

Standard economic theory views fuel and energy as just one set of inputs that is fully substitutable with other inputs, but this is incorrect. Free energy upgrades and organizes all other inputs, and it is a complement to the production process that cannot be created by combining other factors of production. The specific amount of energy needed to produce goods and services is called embodied energy.

If one considers the last 100 years of the US experience, fuel use and economic output are highly correlated. An important measure of fuel efficiency is the ratio of energy use to the gross national product, E/GNP. The E/GNP ratio has fallen by about 42% since 1929. We find that improvements in energy efficiency are due principally to three factors: (1) shifts to higher-quality fuels such as petroleum and primary electricity; (2) shifts in energy use between households and other sectors; and (3) higher fuel prices. Energy quality is by far the dominant factor.

7.5.2 Labor productivity and technological change

Hypothesis 2: A large component of increased labor productivity over the past 70 years has resulted from increasing human labor's ability to perform physical work by empowering workers through increasing quantities of fuel both directly and as embodied in industrial capital equipment and technology.

Economic models generally present technological advances as a means to increase labor and capital productivity. These effects of technological change are measured as a residual after accounting for all tangible factors; energy and natural resources are not considered tangible factors, thus leaving a large residual. From an energy perspective, however, increases in labor productivity are driven by increased fuel use per worker hour. In the pre-1973 period, when fuel prices were falling relative to the price of labor (the wage rate), labor productivity rose as fuel was substituted for labor due to changes in their relative prices. In the post-1973 period, as the price of fuel rose relative to wage rates, the data indicate declining labor productivity.

7.5.3 Energy and inflation

Hypothesis 3: The rising real physical cost of obtaining energy and other resources from the environment is one important factor that causes inflation.

High inflation rates can be explained by the linkages between fuel use and money supply. If the money supply is increased, stimulating demand beyond levels that can be satisfied by existing fuel supplies, then prices will rise. This implies that when the costs of obtaining fuel are high, fiscal and monetary policies may not be successful in stimulating economic growth.

7.5.4 Energy costs and technological change

Hypothesis 4: The energy costs of locating, extracting, and refining fuel and other resources from the environment have increased and will continue to increase despite technical improvements in the extractive sector.

It has been argued that technological innovations for mining low-quality ores can address the problems associated with the depletion of high-quality mineral deposits. Evidence of this is seen in the constant or declining amounts of inputs used per unit output in the extractive sector during this century.

From a physical perspective, however, such a sanguine view of the depletion and scarcity of important natural resources is unwarranted. The extraction of lower-quality ores requires the use of more energy-intensive capital and labor inputs. Over the last few decades, there has been an increase in the direct fuel input per unit of output of fuels and minerals. The present rising energy costs of fuel extraction do not bode well for future exploitation of nonrenewable resources.

The EROIs for natural gas, petroleum, and coal have fallen dramatically over time in the continental United States. In Louisiana, the EROI for natural gas declined from 100:1 in 1970 to 12:1 in 1981, and a similar decline was observed in the petroleum industry. Nationally, the EROI for coal has fallen from 80:1 in the 1960s to 30:1 in 1977. Another indicator of the increasing cost of fuel extraction is the rise in the real dollar value of the mining sector share of real GNP, from 3%–4% over most of this century to about 10% by 1982. Continued economic growth depends on our ability to develop sources of energy with more favorable EROIs.

7.6 Conclusion

Declining EROIs for fuels and increasing energy costs for nonfuel resources will negatively affect economic growth, productivity, inflation, and technological change. To maintain current levels of economic growth and productivity, we must either develop alternative fuel technologies with EROI ratios comparable to those of petroleum today or increase the efficiency of the fuel used to produce economic output.[6]

7.6.1 Circular economy

The Ellen MacArthur Foundation summarizes the ideas behind circular economy: "A circular economy is one that is restorative and regenerative by design and aims to keep products, components, and materials at their highest utility and value at all times, distinguishing between technical and biological cycles. This new economic model seeks to ultimately decouple global economic development from finite resource consumption. A circular economy addresses mounting resource-related challenges for business and economies, and could generate growth, create jobs, and reduce environmental impacts, including carbon emissions. As the call for a new economic model based on systems-thinking grows louder, an unprecedented favorable alignment of technological and social factors today can enable the transition to a circular economy." Furthermore, the foundation states that "the circular

[6] Author's note: The empirical analyses in this article have been enriched and updated. An additional decade of information substantiates the basic conclusions of the article. The interested reader is referred to Robert K Kaufmann, "A Biophysical Analysis of the Energy/GDP Ratio," *Ecological Economics 6* (July 1992): 35–56; and Robert K Kaufmann, "The Relation Between Marginal Product and Price: An Analysis of Energy Markets," *Energy Economics 16* (1994): 145–48 [pp. 211–214].

model of growth, decoupled from the consumption of finite resources and capable of delivering resilient economic systems, is increasingly looked upon as the next wave of development." "The circular economy, by moving much more biological material through the anaerobic digestion or composting process and back into the soil, will reduce the need for replenishment with additional nutrients. Systematic use of available organic waste could help regenerate land and replace chemical fertilizers 2.7 times over." (**Kozo M., Gianpietro M.,** Reconsidering "Circular Economy" In Terms of Irreversible Evolution of Economic Activity and Interplay Between Technosphere and Biosphere, Romanian Journal of Economic Forecasting—XXII (2) 2019, pp. 197–207.

Since AI has penetrated our consciousness these days, we have generated some text on the topics mentioned using ChatGPT:

Entropy is a fundamental concept in both physics and information theory that measures the amount of disorder or randomness in a system Energy intensity refers to the amount of energy required to produce a unit of economic output, such as GDP. Organization and technology are two factors that can influence energy intensity.

Organizational factors can affect energy intensity by influencing the efficiency with which energy is used in production processes. For example, a well-organized company may have better systems in place for managing energy use, such as energy audits, employee training programs, and energy-efficient equipment. Such measures can reduce energy waste and increase the efficiency of production, thereby reducing energy intensity.

Technology can also have a significant impact on energy intensity. Advances in technology can lead to the development of more efficient production processes, such as the use of renewable energy sources or the adoption of energy-efficient equipment. For example, the use of LED lights instead of traditional incandescent bulbs can significantly reduce energy consumption in lighting applications.

Moreover, technology can enable the development of new products and services that require less energy to produce, which can reduce energy intensity at the macroeconomic level. For instance, the use of digital communication technologies, such as video conferencing and email, can reduce the need for business travel, thereby reducing energy consumption.

In summary, energy intensity can be influenced by organizational factors, such as efficient systems for managing energy use, and technological factors, such as the development of more energy-efficient production processes and products. By focusing on these factors, policymakers and businesses can work toward reducing energy intensity and achieving more sustainable economic growth.

Refers to the amount of energy required to produce a unit of economic output, such as GDP. Organization and technology are two factors that can influence energy intensity.

Organizational factors can affect energy intensity by influencing the efficiency with which energy is used in production processes. For example, a well-organized company may have better systems in place for managing energy use, such as energy audits, employee training programs, and energy-efficient equipment. Such measures can reduce energy waste and increase the efficiency of production, thereby reducing energy intensity.

Technology can also have a significant impact on energy intensity. Advances in technology can lead to the development of more efficient production processes, such as the use of renewable energy sources or the adoption of energy-efficient equipment. For example, the use of LED lights instead of traditional incandescent bulbs can significantly reduce energy consumption in lighting applications.

Moreover, technology can enable the development of new products and services that require less energy to produce, which can reduce energy intensity at the macroeconomic level. For instance, the use of digital communication technologies, such as video conferencing and email, can reduce the need for business travel, thereby reducing energy consumption.

In summary, energy intensity can be influenced by organizational factors, such as efficient systems for managing energy use, and technological factors, such as the development of more energy-efficient production processes and products. By focusing on these factors, policymakers and businesses can work toward reducing energy intensity and achieving more sustainable economic growth.

In econophysics, entropy has been used to study complex economic systems, such as financial markets, by analyzing the distribution of returns and volatility in those markets. This approach is based on the idea that economic systems exhibit self-organized criticality, which means that small changes in one part of the system can lead to large-scale effects throughout the system.

One example of the use of entropy in econophysics is the study of stock market crashes. Researchers have found that the distribution of stock returns during a crash follows a power law, which suggests that the market is in a state of high entropy. This high entropy state is associated with a loss of predictability and an increased risk of systemic collapse.

In the context of climate change, entropy has been used to analyze the Earth's climate system and the flow of energy within it. The Earth's climate is a complex system that exhibits high levels of entropy due to its nonlinear dynamics and the presence of feedback loops. Entropy can be used to measure the amount of energy flow and dissipation within the climate system, which is important for understanding the causes and effects of climate change.

One example of the use of entropy in climate change research is the study of atmospheric circulation patterns. Researchers have found that changes in atmospheric circulation can lead to a decrease in the amount of entropy dissipation, which can result in more extreme weather events such as heat waves and droughts. This suggests that entropy can be used as a tool for predicting the impacts of climate change on the Earth's climate system.

Overall, the concept of entropy is a powerful tool for understanding complex systems in both econophysics and climate science. By measuring the amount of disorder or randomness in a system, researchers can gain insights into the underlying dynamics of these systems and make predictions about their future behavior. Energy intensity refers to the amount of energy required to produce a unit of economic output, such as GDP. Organization and technology are two factors that can influence energy intensity.

Organizational factors can affect energy intensity by influencing the efficiency with which energy is used in production processes. For example, a well-organized company may have better systems in place for managing energy use, such as energy audits, employee training programs, and energy-efficient equipment. Such measures can reduce energy waste and increase the efficiency of production, thereby reducing energy intensity.

Technology can also have a significant impact on energy intensity. Advances in technology can lead to the development of more efficient production processes, such as the use of renewable energy sources or the adoption of energy-efficient equipment. For example, the use of LED lights instead of traditional incandescent bulbs can significantly reduce energy consumption in lighting applications.

Moreover, technology can enable the development of new products and services that require less energy to produce, which can reduce energy intensity at the macroeconomic level. For instance, the use of digital communication technologies, such as video conferencing and email, can reduce the need for business travel, thereby reducing energy consumption.

In summary, energy intensity can be influenced by organizational factors, such as efficient systems for managing energy use, and technological factors, such as the development of more energy-efficient production processes and products. By focusing on these factors, policymakers and businesses can work toward reducing energy intensity and achieving more sustainable economic growth.

Generated with ChatGPT.

Reference

Brooks, D.R., Wiley, E.O., 1988. Evolution as Entropy: Toward a Unified Theory of Biology. University of Chicago.

Further reading

Weber, B.H., DePew, D.J., Smith, J.D. (Eds.), 1988. Entropy, Information and Evolution: New Perspectives on Physical and Biological Evolution. MIT Press.

Wicken, J.S., 1987. Evolution, Thermodynamics and Information: Extending the Darwinian Program. Oxford University Press.

See also http://www.fes.uwaterloo.ca/u/jjkay/pubs/.

CHAPTER EIGHT

Gibbs "paradox" in modal multivalued logic of experimenter

Experience is knowledge. All the rest is information.

Albert Einstein.

Chapter 8 introduces an original approach to the so-called Gibbs "paradox" related to the interaction of two interconnected thermodynamic systems based on a modal logic approach (different from the usual binary logic) in terms of the information available to the decision-maker. This chapter stresses that for correct decisions, one should not oversimplify the amount of information needed by decision-makers.

Correlation of the economy and the environment may be approached within the framework of the "Gibbs paradox," where the situation of two separated and one united volume with particles is analyzed.

All papers on the Gibbs paradox that I have seen use a binary logic to describe the level of knowledge of the observer. In practice, this involves not just the case of observing the number of particles in the typical two volumes separated by a membrane. To acquire knowledge, one must interact with the gas and measure the number of particles. To perform the measurement operation, the experimenter (a new name for the operator) creates conditions for measuring. The result of one measurement operation provides the possibility that the measurement is true or false. To obtain more knowledge about the state of truth of the final result, the experimenter must successively repeat the conditions and thus repeat the measurement. The new result will increase the level of knowledge of the experimenter until the experimenter knows all the particles in the two volumes. Only after reaching that state may one apply the calculation of entropy variation used in describing the Gibbs entropy variation based on a binary logic—this being an extreme case of the evolution of the state of knowledge of the experimenter.

Climate Change and Circular Economics
ISBN: 978-0-443-29969-8
https://doi.org/10.1016/B978-0-443-29969-8.00013-6

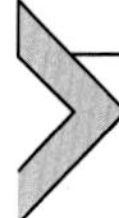

8.1 Measuring technological information and entropy

The preceding comments on resource allocation lead in a wider context to the problem of creating technological information and further to entropy as a measure of economic transitions, taken in the sense of Georgescu–Roegen.

In the case of correlation between technological information and GDP productions described above, we encounter the possibility of a chaotic regime with several trajectories of evolution confined to a "strange attractor" (of the Hénon type), so it is useful to analyze the specific parameters for such nonlinear dynamics.

The approach taken here involves escort distributions and Rényi information measures, but we first introduce a measure of information gain that is more appropriate, in our view, for economic processes that are strongly anchored in experimental reality and have an underlying modal logic.

In economic processes like the one described above, the underlying logic is not always bivalent. By recreating the conditions for technological knowledge generation, information is gained on the state of the truthfulness or falseness of the technological sequences under research and implementation. Let us consider not only two states, i.e., true (functions, generates more GDP, is more efficient, etc.) and false (does not work, does not generate more GDP, is less efficient, etc.) of the technological sequence of events under consideration but also any number of possible values between true and false. Each intermediate value can be expressed as a combined measure of the measure of true and the measure of false technological sequences (Fig. 8.1).

From the point of view of research and implementation, we may define the specific measures of the two states as the probabilities that technological sequences will be true (in the sense above) defining, for true, the normalized probabilities of the technological sequence i as $P_{it} = (p_i)^{\beta}/\Sigma(p_j)^{\beta}$, where p_i are not zero. Here the distributions p are given by the observed relative frequencies of the technological experimental tests. In the same manner, we define the measure of false P_{if} for the sequence where instead of p we take $1-p$. The two measures result from research and development activity that separates working technological events from those that do not work in the process to create useful technologies. In fact, the research activity creates conditions for testing the technological sequences and gains technological information from test outcomes. The gain in information is described below.

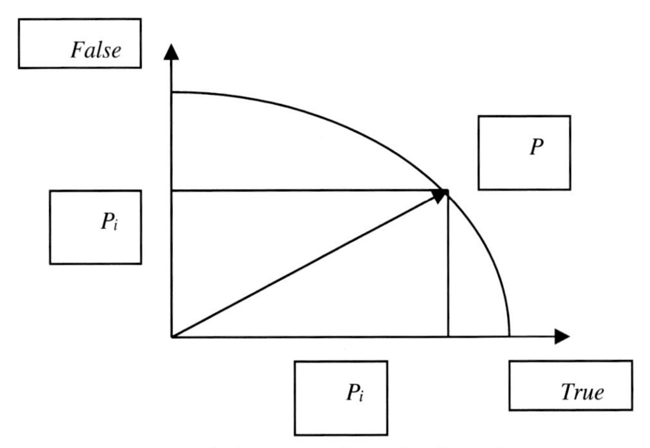

Figure 8.1 Knowledge vector for technological sequence.

We may define a vector of the state of knowledge for each experimental stage in technological development and associate it with a measure that results from typical vector calculus as $P^2 = P_{it}^2 + P_{if}^2$. Further, we consider that research and development repeats the conditions for testing technologies, and this adds a third dimension to the first two (true and false) marking the passage of time. This dimension we associate with a measure of time that results from the frequency of repeating the tests, denoted by iP_0 (with $i = \sqrt{(-1)}$ marking a rotation by 90 degrees in the complex space). After each development test, the vector of knowledge changes by gaining technological information, so we may describe the change in information in the knowledge vector as a space defined as a Minkowski space in physics. As shown in Purica (1990), the gain in information is described by a Lorentz transformation. The resulting vector has a magnitude $P^2 = P_{it}^2 + P_{if}^2 - P_0^2$. Those familiar with relativity theory will recognize the specific three-dimensional cones that separate the space into three regions. These have a specific meaning in our interpretation. Inside the cone, we have technological information that is generated from tests and keeps a certain incertitude described by $P_0^2 > P_{it}^2 + P_{if}^2$. On the cone, the technological information is coherent and leads to implementable technological sequences, while outside the cone, $P_0^2 < P_{it}^2 + P_{if}^2$ (this would be an unreal situation when the frequency of technological sequence tests is lower than the number of tests on technological knowledge) (Fig. 8.2).

We further analyze entropy variation as it evolves with the level of knowledge of the experimenter. Zero knowledge and complete knowledge

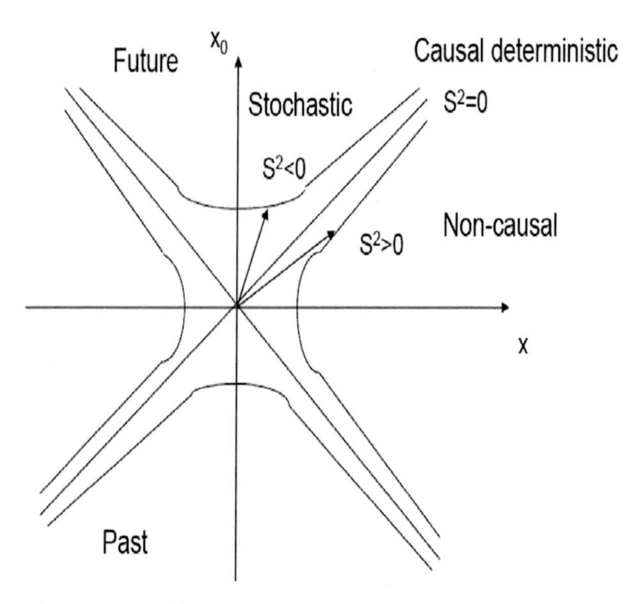

The pass from a state of knowledge to another is described by a Lorentz transform in the Minkovski space. The evolution of the state of knowledge is associated with a 'trajectory of knowledge'.

Figure 8.2 Evolution of the state of knowledge.

are limited cases of this analysis. Going along with Purica (1977, 1990, 2010), the description of the measurement—modal space—may be represented by a Minkowski space where the time axis shows the repeated measurement conditions while the other two axes give the truth and false statuses of the results. Each measurement determines a result with a given possibility of being true or false. The knowledge vector description has a measure whose square value is the sum of the squares of the measure of truth and the measure of false. Considering the time axis complex, the vector of knowledge has a measure squared equal to the sum of the squares of the measures of the truth and false values minus the square of the time axis measure (time axis is complex). If we consider the angle of projection of the knowledge vector on the plane of the truth and time axes, the cosine of the angle with the truth axis equals the hyperbolic tangent of the angle between the knowledge vector and the time axis. The value of the hyperbolic tangent is the square root of the ratio of the number of true results and the

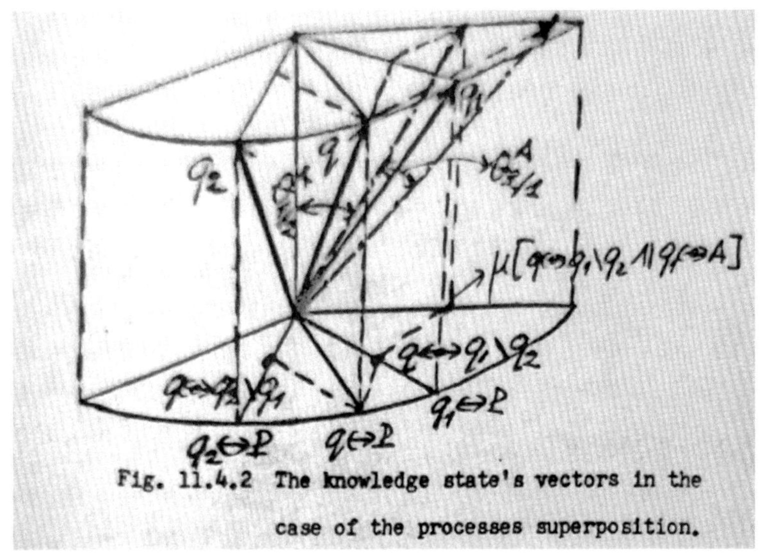

Figure 8.3 Space of superposition of two processes. *Source: Purica I. (1977). The Laws of Modal Thought. Editura Centrul National de Fizica, Bucharest, The Laws of Modal Thought.*

number of measurements. Fig. 8.3 represents the situation of two cases superposed as separate volumes but also combined.

The calculation of the value of entropy variation from the case of two separate volumes to their combination, considering the knowledge level of the experimenter, uses the square of the hyperbolic tangent of the knowledge angle as a measure of the probability of the state in each volume:

$$S = -\,th^2\theta_1\; ln\left(th^2\theta_1\right) - th^2\theta_2\; ln\left(th^2\theta_2\right)$$

where the index relates to the volumes, and $q_i = a\mathrm{th}(\mathrm{sqrt}(n_i/n)$, $n = n_1 + n_2$, n_i is the number of particles in each volume. If the number of particles is measured completely, we obtain the limit case where the same number of particles in the two volumes provides a value of $S = \ln 2$. This is the limit case when the knowledge level is maximum after all particles have been detected through measurement (Fig. 8.4).

The combination of the state of knowledge resulting from measurements of the number of particles in the two volumes is given by

$$th(\theta_1 + \theta_2) = (th\;\; \theta_1 + th\;\; \theta_2)\,/\,(1 + th\;\; \theta_1 \cdot th\;\; \theta_2)$$

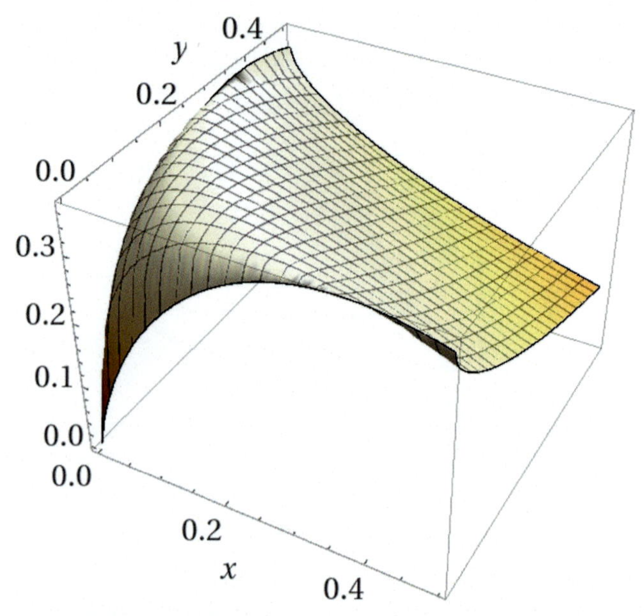

Figure 8.4 State of knowledge for two processes.

Thus, the evaluation of the difference in entropy from the state of the separate volumes to the one where they are combined is given by

$$S = -th^2(\theta_1+\theta_2)ln\left(th^2(\theta_1+\theta_2)\right) - \left(-th^2\theta_1\ ln\left(th^2\theta_1\right) - th^2\theta_2\ ln\left(th^2\theta_2\right)\right)$$

The representation of the components of the above formula in the space (θ_1,θ_2,S) is given below. It should be noted that complete knowledge may be reached at various trajectories on the surface represented by S (Fig. 8.5).

Moreover, the difference in the functions above, represented below, shows that the greater the knowledge about the system's total volume compared with the situation of separate volumes, the lower the entropy passing into the negative values, although at a low knowledge level, the value is positive. This entropy evolution space at various knowledge levels of the experimenter may indicate that complete knowledge is associated with negative entropy (Fig. 8.6).

To conclude the above reasoning, we note that in the space of possible values determined by measurement, there can be deterministic and statistical states, as well as those associated with quantum processes. The evolution of

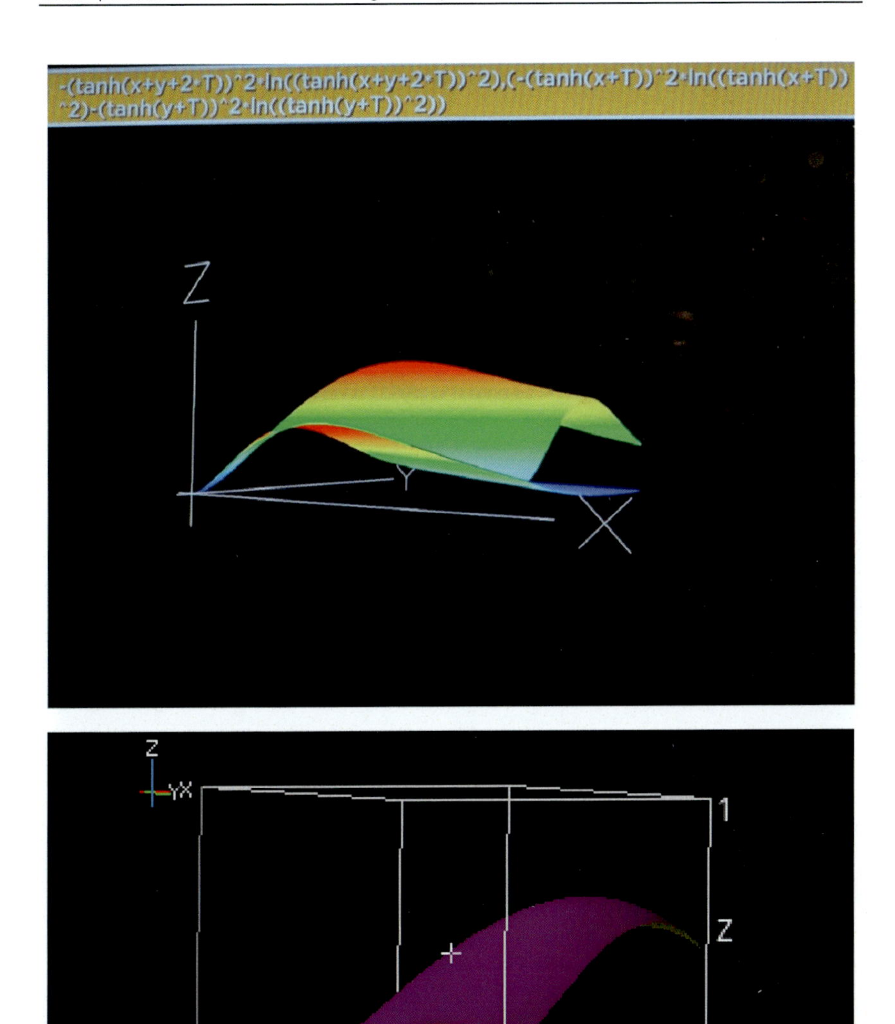

$Z=S$, $\theta_1=x$, $\theta_2=y$

Figure 8.5 Entropy for the case of isolated and combined processes.

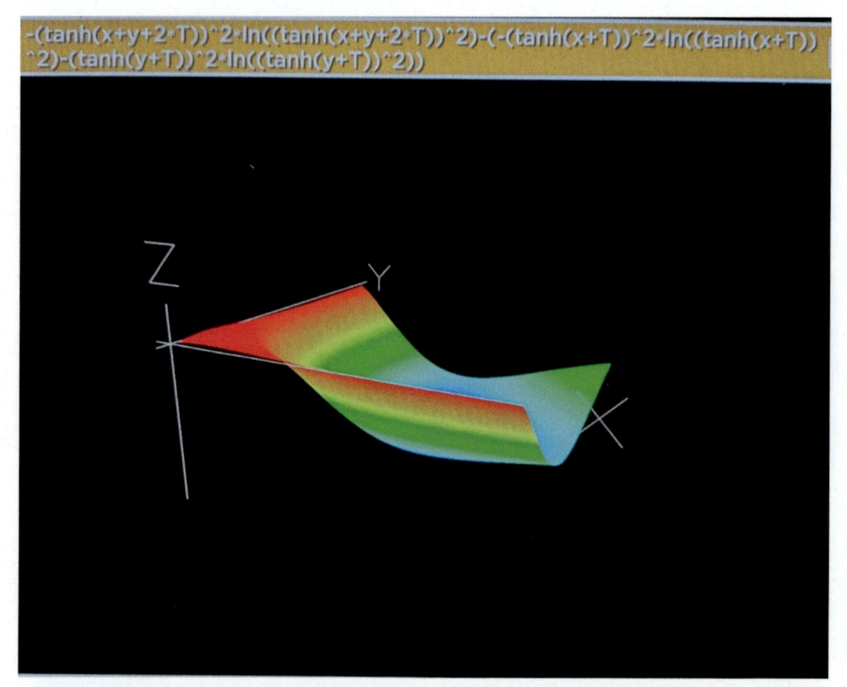

Figure 8.6 State of knowledge and entropy for combined system.

the knowledge vector is represented by a Lorentz transformation in the specific space, suggesting that quantum bit combinations may also be described within this multivalued logic approach.

References

Purica, I., 1977. (The Laws of Modal Thought). Editura Centrul National de Fizica, Bucharest.

Purica, I., 1990. Legile Gândirii Modale, Editura Academiei, Bucuresti.

Purica, I., 2010. Nonlinear Models for Economic Decision Processed. Imperial College Press, London.

CHAPTER NINE

United Nations sustainable development goals

It is our choice of good or evil that determines our character, not our opinion about good or evil.

Aristotle.

This ninth chapter presents the Sustainable Development Goals (SDGs) defined by the United Nations (UN) as proof of the complexity of sustainable development in human society, again stressing the need for a complex vision to address this complex process.

The UN's Sustainable Development Strategy provides a comprehensive framework for an integrated vision of development. It includes the introduction of 17 SDGs covering a multitude of activities.

On September 25, 2015, the 193 countries of the UN General Assembly adopted the 2030 Development Agenda, titled "Transforming our world: the 2030 Agenda for Sustainable Development." The agenda has 92 paragraphs, with paragraph 59 outlining the 17 SDGs along with the 169 targets and 232 indicators associated with them.

The UN-led process involved its 193 member states and global civil society. The resolution is a broad intergovernmental agreement that acts as the Post-2015 Development Agenda. The SDGs build on the principles agreed upon in Resolution A/RES/66/288, titled "The Future We Want." This nonbinding document was released as a result of the Rio+20 Conference held in 2012.

The following is the full list of the 17 SDGs with a brief description of each:

Goal 1

End poverty in all its forms everywhere.

Goal 2

End hunger, achieve food security and improved nutrition and promote sustainable agriculture.

Goal 3

Ensure healthy lives and promote well-being for all at all ages.

Climate Change and Circular Economics
ISBN: 978-0-443-29969-8
https://doi.org/10.1016/B978-0-443-29969-8.00011-2

Goal 4

Ensure inclusive and equitable quality education and promote lifelong learning opportunities for all.

Goal 5

Achieve gender equality and empower all women and girls.

Goal 6

Ensure availability and sustainable management of water and sanitation for all.

Goal 7

Ensure access to affordable, reliable, sustainable and modern energy for all.

Goal 8

Promote sustained, inclusive and sustainable economic growth, full and productive employment and decent work for all.

Goal 9

Build resilient infrastructure, promote inclusive and sustainable industrialization and foster innovation.

Goal 10

Reduce inequality within and among countries.

Goal 11

Make cities and human settlements inclusive, safe, resilient and sustainable.

Goal 12

Ensure sustainable consumption and production patterns.

Goal 13

Take urgent action to combat climate change and its impacts.

Goal 14

Conserve and sustainably use the oceans, seas and marine resources for sustainable development.

Goal 15

Protect, restore and promote sustainable use of terrestrial ecosystems, sustainably manage forests, combat desertification, and halt and reverse land degradation and halt biodiversity loss.

Goal 16

Promote peaceful and inclusive societies for sustainable development, provide access to justice for all and build effective, accountable and inclusive institutions at all levels.

Goal 17

Strengthen the means of implementation and revitalize the global partnership for sustainable development.

By Own work using: File: Sustainable Development Goals.png and PDF infographic from un.org—Traced from File: Sustainable Development Goals.png, from un.org, Public Domain, https://commons.wikimedia.org/w/index.php?curid=81280117

The 2030 horizon has triggered the elaboration of national strategies now being executed by various UN member countries.

Along a different line, consistent activity to standardize these activities is reflected in ISO standards such as 37100 and 238 W3. The advantage of having standardized activities creates, besides the possibility of benchmarking and transferring knowledge, a uniform environment for implementing research results.

One example that we may give, which was recognized by the UN, is Romania. This country has organized a dedicated department for sustainable development at the level of the prime minister. The department's role is not only to perform the country's 2030 strategy but also to identify the basic indicators for each SDG and monitor the implementation of the associated strategy.

Moreover, specific job profiles were created for sustainable development experts who will be associated with local administration and ministries to provide the needed knowledge for implementing each SDG. The National Institute of Statistics follows the indicators for each SDG and periodically contributes to the National Voluntary Report on SDGs at the implementation stage.

A database has been created, and each ministry contributes specific data. The introduction of courses on SDGs is envisaged for universities, and public dissemination campaigns are conducted periodically.

A national Advisory Council for Sustainable Development was set up, gathering recognized experts from each SDG domain.

As a potentially relevant example of the national strategy, we present an excerpt of the energy component (SDG 7).

SDG 7 Ensure access to affordable, reliable, sustainable and modern energy for all.

Global energy demand is constantly growing, and only by encouraging energy efficiency and promoting renewable energy can current needs and those of future generations be met. The energy sector plays an essential role in Romania's development through its profound influence on the competitiveness of its economy, quality of life, and the environment. To meet consumer expectations in the long term, the Romanian energy sector needs to become more robust from an economic point of view, more advanced and more flexible from a technological point of view, and more environmentally friendly.

9.1 Energy infrastructure

Boasting a value of approximately 30 billion euros and employing some 80,000 people in 2017, the energy sector is a large and strategically important part of the Romanian economy. It is also the main emitter of greenhouse gases (GHGs) and as such plays a central role in reducing global warming in terms of both gradually shifting away from fossil fuels toward sources with low GHG emissions, particularly solar and wind, and increasing the energy efficiency of buildings and vehicles and switching to electric transportation technology.

9.2 Energy security

Romania is fairly well placed within a regional and European context when it comes to energy (https://ec.europa.eu/energy/en/topics/energy-strategy-and-energy-union/clean-energy-all-europeans) security. Its rate of energy independence in 2016 was 78.4%, with 80.3% for coal (including coke), 33.1% for crude oil, and 86.4% for natural gas. The advantage of Romania having internal primary energy sources is also reflected in the country's balanced energy mix, comprising coal, nuclear, natural gas, and renewables hydro, wind, solar, and biomass. However, the situation in the international energy market is dynamic, and technological developments can have unforeseen effects on energy markets. Ambitious policies at the European level in the field of energy and climate change51 have focused on reducing GHGs, expanding the share of renewable sources, and shifting public attitudes toward "clean energy," all of which will influence investment behavior in the energy sector as well as patterns of energy consumption.

9.3 Access to energy

Current priorities in the energy sector are focused on correcting certain dysfunctionalities and bringing Romania in line with EU averages. These priorities include expanding electricity connectivity and natural gas supply networks at regional and European levels to eliminate congestion, gradually decarbonizing, and electrifying final energy consumption in all sectors of the economy including the residential environment, replacing physically and conceptually outdated technologies, and adopting an integrated approach to policies in energy and other sectors, while respecting the obligation to protect the environment and limit the consequences of climate change. In terms of access to energy, according to Eurostat data for 2016, 38.8% of the population in Romania, i.e., 7,6000,000 people, were susceptible to social exclusion. In absolute terms, during the period 2008–16, 1,420,000 people were removed from the category of people at risk of poverty or social exclusion.

9.4 Renewable energy and energy efficiency

Romania achieved EU targets for 2020 in terms of the contribution of energy policies to reduce the impact of climate change in advance. As part of its commitment to reduce GHG emissions by 20% relative to 1990 levels, in 2012 Romania achieved a reduction of 47.96% compared with the EU average of 82.14%. In 2016, Romania saw an increase of 25.03% in its share of energy derived from renewable sources in gross consumption and a reduction in energy consumption of 41.6% compared with the 20% it had committed to. To maintain the share of renewable sources committed to by Romania, a series of legislative changes were introduced as Law 220/2008 to establish a system to promote renewable energy production, with subsequent modifications and additions, including integrating small producers into the national energy system. Romania is implementing the National Energy Efficiency Action Plan 52 Tons of oil equivalent 53 Progress Report on the Meeting of National Energy Efficiency Targets, ANRE, April 2018 54 Ibid approved in 2015. Work is currently underway on the Integrated National Energy and Climate Change Plan, which is based on the Draft Energy Governance Regulation that contains the targets and measures that must be implemented at the EU level in this field. Primary energy consumption is a key indicator in monitoring the progress made by the EU and each member state to achieve the targets established by Directive (2012)/27/EU. In 2016, Romania had the lowest level of primary energy consumption of all 28 EU member states, at 1582 tons of oil equivalent per capita, 52 which was almost half the EU average for the same year, at 2.997 toe per capita. The level of primary energy consumption per capita for 2016 nonetheless represents a reduction of 8.3% from 2011. Romanian energy efficiency policies made an important contribution to this reduction. While final energy consumption increased by 1.8% in 2016 compared with the previous year, with the value of GDP also being 4.8% higher, 54 the level of consumption for the 2014–16 period was below that of 2011–12. The household sector accounted for the greatest share of final energy consumption, at 34.5% in 2011 and 33.2% in 2016. This suggests that efforts in this sector must focus on increasing energy efficiency through programs for the thermal insulation of residential apartment buildings, the labeling of electrical goods, and the Green Homes program. The amount of energy saved through the thermal rehabilitation of buildings as part of the Regional Operational Program for 2007–13 was 348 GWh/year, with 41,311 apartments rehabilitated. Further energy savings were achieved through the Green Homes program for natural persons for the installation of heating systems based on renewable energy and implemented by the Environment Fund Administration, which saw the installation of 40,000 systems worth approx. 250,000,000 lei. The renovation of buildings is an excellent opportunity for the sustainable modernization of existing architecture, resulting in multiple benefits for households, businesses, and the public sector. A strategic integrated approach in this respect would stimulate the market positively and differently from the current fragmented initiatives. Through the revision of Directive (2010)/31/EU of the European Parliament and the Council of May 19, 2010, on the energy performance of buildings, the European Commission encourages the use of innovative and smart technologies in buildings. Two-thirds of all buildings in the EU were built before the development of building regulations, and the percentage of these that have been renovated is approximately 1%. 55 Progress Report on the Meeting of National Energy Efficiency Targets, ANRE, April 2018 56 Energy intensity is a measure of gross inland consumption of energy related to the national economy (i.e., the amount of energy necessary to produce one unit of GDP) 57 Purchasing power parity The act of authorizing energy auditors/certifying energy managers helps promote and develop a system for the provision of audits that can highlight potential energy savings for the final energy consumer. The corresponding number of energy auditors certified each year indicates the openness of the market to energy services. At the end of 2017, there were 441 energy managers, 207 energy auditors who were natural persons, 72 energy auditors who were legal persons (of whom 19 were registered sole traders (PFA)), and 71 providers of approved energy services (of whom 20 were registered sole traders (PFA)).55 In 2016, the energy productivity indicator for Romania had a value of EUR 10.30 PPP/kgoe, which was higher than the EU-28 average of EUR 9.1 PPP/kgoe.56 This places Romania in sixth place among EU member states.57 The system for the promotion of electricity produced from renewable sources through green certificates has been in place since 2005. However, Romania is still at the stage of adapting to the green economy/green energy.

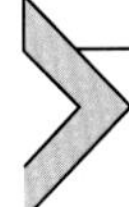

9.5 Targets for 2030

- Expand electricity and gas distribution networks to ensure household consumer, industrial, and commercial access to safe sources of energy at acceptable prices.
- Ensure the cybersecurity of platforms for monitoring the production, transport, and supply networks of electricity and natural gas.
- Decouple economic growth from the processes of resource depletion and environmental degradation by substantially boosting energy efficiency (by a minimum of 27% compared with the status quo) and the extensive use of the EU Emission Trading Scheme under stable and predictable market conditions.
- Increase the share of renewable energy and low-carbon fuel used in the transport sector (electric vehicles), including alternative fuels.
- Ensure a stable and transparent regulatory framework in the field of energy efficiency to attract investment.

- Strategically support the share of electricity in total household, industrial, and transport consumption by establishing performance standards for facilities and equipment.

The strategy is periodically reviewed to include the evolution of each SDG domain that could not be predicted when initially elaborated. For example, the change in the energy paradigm induced by the war in Ukraine (at the border of Romania) has an impact on system operations and accelerates the use of existing energy resources in the country.

A suggestion for the future in addressing SDG evolution would be to consider the matrix of intercorrelation among the SDGs. That would make the potential nonlinear effects more relevant for decision-making.

CHAPTER TEN

Final thoughts on approaching the future

I think; therefore I am.

Rene Descartes.

This last chapter synthesizes the book's conclusions and presents evolutionary scenarios for a sustainable future with insights on potential drawbacks that could delay achievement of the final goal of a coherent and durable relationship between human society and the environment.

Looking at the evolution of emissions, one may wrongly conclude that emissions decrease only in times of crisis. Economic crises include the reform of Eastern European economies during the 1990s, the financial crises in 2008, the pandemic crises in 2020, and the war crises in 2022. Considering emissions as the only indicator for climate change control results in a discrepancy between the desired evolutionary future (i.e., 1.5°C with net zero emissions) and the reality of the past evolution. Moreover, neglecting the innovation effect of new technologies within all domains may lead to decisions that neglect important resources used by society, like abandoning coal use instead of developing and implementing clean coal technologies.

To achieve a coherent relationship with the environment, human society must become a closed system in thermodynamics terms. This is explained in the beginning, and to sustain the behavior of existing economies facing a typical evolution and the interdependence of specific parameters, a model is devised to analyze responses to changes in parameters. The results point to the fact that long-term economic behavior may be simulated to match that observed in reality for different economies and various relations to climate change that occur naturally from the model equations. Based on the long-term evolution of energy intensity recorded for various economies, a method to calculate the contribution to the increase of the environmental temperature was developed based on the irreversible thermodynamics approach using the Gibbs equation.

Several conclusions can be drawn from this approach:

(i) The value of the temperature increase in each economy may be determined and compared among economies.

Climate Change and Circular Economics
ISBN: 978-0-443-29969-8
https://doi.org/10.1016/B978-0-443-29969-8.00008-2

(ii) The determined speed of the temperature increase provides an important means of comparing it with the speed of environmental recovery (resilience), creating a basis for analyzing the dynamics of the potential impact on climate change. Temperature increases for newly emergent economies are contributed at greater speeds (i.e., higher growth rates) and are therefore likely to have a greater impact on environmental damage.

(iii) The approach underlines the importance of resource recycling to reduce temperature increases. This opens the way to new thinking in terms of strategies that accentuate the principles of a circular economy. Development should enhance new technology innovation to bring about new resources and recycle the already used ones, be they material or energy resources. A selection of present-day material resources used in economies and existing and future technologies is presented for a perspective on the orientation of research and innovation efforts. The sequence of innovation cycles is presented, with the fast economic cycles that are shortening Kondratiev cycles are analyzed as an economic case example correlated with the shortening of innovation cycles.

To check the existence of a climate change effect, a big data evaluation is performed for Romania, resulting in maps of potential risks from various climate change events (floods, drought, snow, and freezing). The total resulting risks suggest a potential insurance policy, with different rates for each region under risk. A similar analysis is performed for Italy related to earthquake and landslide risks. These risk evaluations allow a calculation of the impact on gas networks of two respective countries that provides important support for decisions about investments in security improvements to critical gas infrastructure. This analysis may be replicated for other critical infrastructures, such as electrical energy transport networks, roads, and railways.

Considering the partitioning of agricultural land between biofuel crops for cars and food crops for persons is an important decision case that is analyzed for the EU, the USA, and China. The calculation of the optimal partition is based on an energy conservation approach to the energy needed for car transportation and energy needed to meet the daily needs of food for people. The results provide useful information related to the possibility of using biofuels or considering other technologies such as electric vehicles.

Since electric vehicles are possible because of the development of batteries, a presentation of potential energy-storage technologies is also provided.

Moreover, another field of study in energy brings important future developments—the direct conversion of energy technologies. The primary technological ideas in that area are also presented.

A selective list of technologies is detailed, with special attention given to reviving nuclear power technologies. Small modular reactor designs are presented in synthesis with comments on future applications of such technologies in space and remote places.

The recent breakthrough of fusion power in different configurations is mentioned along with thorium reactor technology. It becomes clear that if we desire a massive reduction in emissions in the economy, nuclear technology is the primary existing technology that could achieve such a result quickly enough to avoid crossing the environmental time limits for recovery.

The time constants of technological penetration are also stressed in connection to the acceptance of new technologies that may change traditional values and habits of living.

The perception of economic evolution, due to its complexity, requires adequate integrated indicators that encompass the single-value indicators used to assess development. It is clear that certain indicators will have increasing values that show progress, while others will have decreasing values showing progress. The proper combination of these indicators will provide single values to mark the evolution of an integrated view on economies.

To underline the complexity of the use of thermodynamics and specifically entropy considerations in the economy that are obviously also connected to energy, a selection was made of various opinions related to this topic. The various points of view and approaches again show the matter's complexity. As a test, a brief result was added of a short piece of text generated with ChatGPT artificial intelligence software.

The many facets of the sustainable development vision for human society are best represented by the Sustainable Development Goals (SDGs) in the UN 2030 Agenda for Sustainable Development. A presentation is included of these SDGs together with a strategy for sustainable development and its implementation measures.

Since thermodynamics is used to describe the correlation of the two systems of nature and human society, a solution is presented of the so-called Gibbs paradox for the case of two separate systems that are also united. The approach considers modal multivalued logic of possibilities that is different from the usual binary logic applied to describe the experimenter's perception of the process. The result points to the importance of the decision-maker having complete information about the system in order to make optimal decisions.

The considerations above are based on the idea that human society must have an in-depth understanding of its relationship to the environment. This is a complex correlation that impacts both humans and the environment. The final goal of this relation must be only exchanging energy with the environment. Resources will be recycled and transformed by new technologies stemming from successive innovation cycles that will also bring change to the living values and traditions of human society regardless of where they may be on Earth, Mars, or other places. Along this line, we should not stop using resources solely because they are polluting (coal is a good example in energy and transportation as a source of local pollution) but rather replace polluting technologies with nonpolluting ones (coal with clean coal technologies, and transportation with electric vehicles and fuel cells).

In addition, some cycles must be managed, a good example being the cement industry with its large volume of emissions. These emissions increase the temperature, leading to increases in the level of seawater. Protection against this process is provided by dams made with more cement and hence more emissions, thus amplifying the cycle.

Along a different line, traditional human behavior must be accounted for if new technologies are to be implemented. If one wants to reduce energy consumption, a technology that may lead to this result is infrared vision devices. We need light because we do not see in the dark. With infrared devices for everyone, the need for public lighting at night time and lights in the buildings would cease. But are we ready to renounce our present habits as fast as we accepted mobile phones, or will tradition prevail over the benefits of new technologies, lengthening the time constants of change?

From the dawn of humanity there has been a continuous battle for resources. The best illustration I have is the beginning of the movie *2001: A Space Odyssey*, where the primitive humans were fighting for a pond of water, and technological knowledge (regardless of where it came from) determined who prevailed. Now, at the level of the whole planet, humans have reached limits in their interactions with the environment, and if these limits are not mastered, then 65 million years from now, a civilization of ants will wonder why the "dinosaurs" disappeared.

A simple indicator approach for such a complex problem may create a state of public behavior similar to that of a religion. I was once asked by a friend if I *believe* in climate change; my answer was that I am still conducting research.

The issue is that due to simple thinking, development decisions may not consider the potential drawbacks and could result in unwanted evolution. Moreover, such decisions may not account for dynamic environmental time constants and produce effects that seem favorable in the short term based on their amplitude and specifics but have irreversible effects on the environment in the long run.

Decisions should be based on complete information (or as complete as possible) for decision-makers. The importance of acquiring information is screened by the simple approach, while the understanding of the process of making optimal decisions is incomplete with potentially damaging results. This is a strong justification for continuous research and innovation activity supported by appropriate investment structures that should attract both private and governmental money.

It is desirable not to embark on a policy of repeated crises of all sorts just because during crises emissions are decreasing in response to economic and social disorganization or changes in living habits. Innovation and resource management may be a solution that brings about desired effects in relation to nature.

Prof. Ionut Purica

Index

Note: 'Page numbers followed by *f* indicate figures, *t* indicate tables, and *b* indicate boxes.'

F

G

H

I

Printed in the United States
by Baker & Taylor Publisher Services